Potential Techniques for the Assessment of Joints in Water Distribution Pipelines

Awwa Research Foundation

About the Awwa Research Foundation

The Awwa Research Foundation (AwwaRF) is a member-supported, international, nonprofit organization that sponsors research to enable water utilities, public health agencies, and other professionals to provide safe and affordable drinking water to consumers.

The Foundation's mission is to advance the science of water to improve the quality of life. To achieve this mission, the Foundation sponsors studies on all aspects of drinking water, including supply and resources, treatment, monitoring and analysis, distribution, management, and health effects. Funding for research is provided primarily by subscription payments from approximately 1,000 utilities, consulting firms, and manufacturers in North America and abroad. Additional funding comes from collaborative partnerships with other national and international organizations, allowing for resources to be leveraged, expertise to be shared, and broad-based knowledge to be developed and disseminated. Government funding serves as a third source of research dollars.

From its headquarters in Denver, Colorado, the Foundation's staff directs and supports the efforts of more than 800 volunteers who serve on the board of trustees and various committees. These volunteers represent many facets of the water industry, and contribute their expertise to select and monitor research studies that benefit the entire drinking water community.

The results of research are disseminated through a number of channels, including reports, the Web site, conferences, and periodicals.

For subscribers, the Foundation serves as a cooperative program in which water suppliers unite to pool their resources. By applying Foundation research findings, these water suppliers can save substantial costs and stay on the leading edge of drinking water science and technology. Since its inception, AwwaRF has supplied the water community with more than $300 million in applied research.

More information about the Foundation and how to become a subscriber is available on the Web at **www.awwarf.org.**

Potential Techniques for the Assessment of Joints in Water Distribution Pipelines

Prepared by:
Christopher Reed
NWTC, Durham, UK, DH1 5FJ

Alastair J. Robinson
Advanced Engineering Solutions Ltd., Northumberland, UK NE23 9WF

and

David Smart
Issquared Ltd., Sunderland, UK SR5 2TA

Sponsored by:
Awwa Research Foundation
6666 West Quincy Avenue, Denver, CO 80235-3098

Published by:

Awwa Research Foundation

American Water Works Association

IWA Publishing

DISCLAIMER

This study was funded by the Awwa Research Foundation (AwwaRF). AwwaRF assumes no responsibility for the content of the research study reported in this publication or for the opinions or statements of fact expressed in the report. The mention of trade names for commercial products does not represent or imply the approval or endorsement of AwwaRF. This report is presented solely for informational purposes.

CONTENTS

TABLES

FIGURES

FOREWORD

The Awwa Research Foundation is a non-profit corporation that is dedicated to the implementation of a research effort to help utilities respond to regulatory requirements and traditional high-priority concerns of the industry. The research agenda is developed through a process of consultation with subscribers and drinking water professionals. Under the umbrella of a Strategic Research Plan, the Research Advisory Council prioritizes the suggested projects based upon current and future needs, applicability, and past work; the recommendations are forwarded to the Board of Trustees for final selection. The foundation also sponsors research projects through the unsolicited proposal process; the Collaborative Research, Research Applications, and Tailored Collaboration programs; and various joint research efforts with organisations such as the U.S. Environmental Protection Agency, the U.S. Bureau of Reclamation, and the Association of California Water Agencies.

This publication is a result of one of these sponsored studies, and it is hoped that its findings will be applied in communities throughout the world. The following report serves not only as a means of communicating the results of the water industry's centralized research programme but also as a tool to enlist the further support of the non-member utilities and individuals.

Projects are managed closely from their inception to the final report by the foundation's staff and large cadre of volunteers who willingly contribute their time and expertise. The foundation serves a planning and management function and awards contracts to other institutions such as water utilities, universities, and engineering firms. The funding for this research effort comes primarily from the Subscription Program, through which water utilities subscribe to the research program and make an annual payment proportionate to the volume of water they deliver and consultants and manufacturers subscribe based on annual billings. The programme offers a cost-effective and fair method for funding research in the public interest.

A broad spectrum of water supply issues is addressed by the foundation's research agenda: resources, treatment and operations, distribution and storage, water quality and analysis, toxicology, economics and management. The ultimate purpose of the co-ordinated effort is to assist water suppliers to provide the highest possible quality of water economically and reliably. The true benefits are realized when the results are implemented at the utility level. The foundation's trustees are pleased to offer this publication as a contribution toward that end.

Walter J. Bishop
Chair, Board of Trustees
Awwa Research Foundation

Robert C. Renner, P.E.
Executive Director
Awwa Research Foundation

ACKNOWLEDGMENTS

The authors would like to acknowledge the financial support of the Awwa Research Foundation, and the other companies and utilities involved, without whose contributions, the project would not have been possible.

The research has been guided by an experienced Project Advisory Committee (PAC), who provided valuable guidance, expertise and suggestions. The authors wish to thank these individuals for their time and effort:

Tony Woodward, Thames Water Utilities, UK
Balvant Rajani, National Research Council, Canada
Kanwal Oberoi, Charleston, South Carolina

A number of water utilities offered their time, staff expertise, and facilities for case studies and field investigations. The authors gratefully acknowledge these efforts:

Essex and Suffolk Water, Chelmsford, UK
Northumbrian Water Limited, Durham, UK
United Water Services, New Jersey
Calgary Waterworks, Canada
East Bay MUD, California

The authors would also like to thank Mike Royer (USEPA) for his valuable input in finalising the document and the Awwa Research Foundation program manager Djanette Khiari for all of her help and assistance throughout the duration of the project.

EXECUTIVE SUMMARY

INTRODUCTION TO PROJECT

The objective of this research project was to identify and document key problems associated with the failure of joints in water distribution pipelines and to investigate and report on the performance of existing and emerging techniques for the location, condition assessment and repair of these joints. Consideration has also been given as to how the information relating to testing and condition assessment can be best employed by the utility organisations. A decision support tool has been developed to assist with the selection of appropriate technologies.

BACKGROUND

Buried pipes, valves and fittings can account for up to 70% of the total value of a Water Company asset. Failure of these assets can result in loss of supply to customers, major damage to public property and pose a severe risk of local flooding. Typical pipe failures can be classified as either pipe joint or pipe barrel failure. These failures can result from poor pipe laying procedures, ground movement loading, environmental deterioration, and internal pressure surges.

Pipe barrel failures have been the subject of considerable investigation over a number of years, but many existing distribution networks also have deteriorating joints that reduce the serviceability of the pipeline. Joints are recognised as an important and integral part of any water distribution network, and leaking joints are a common occurrence, but as these failures are not always catastrophic, they often remain undetected or are not acted upon. The detectable failures are observed as leaks or pipe breaks at joints, and may result in the deterioration of water quality, or in extreme cases, loss of supply to the customers. Mechanical joints are particularly problematic due to the complexity of sealing requirements, the need for some rotational capability and perhaps thrust restraint.

Examples of specific types of joint failure could include degradation of natural rubber gaskets and seals, breakage of contact surfaces within rigid lead joints, and corrosion and expansion of sulphur-based jointing compounds. Many existing pipe networks are thought to have deteriorating joints that can reduce the serviceability of the pipeline. These serviceability failures are observed as leaks or pipe breaks at joints, and may result in water quality deterioration.

The primary objectives of this study have been to identify and document the key problems associated with the failure of joints in pipelines and to evaluate and report upon the performance of techniques that could be utilised for the inspection and condition assessment of a wide variety of joint types. The study has also identified and investigated potential methods to repair joints without the expense and disruption resulting from open-trench excavation.

RESEARCH APPROACH

In order to appreciate which joints give rise to most of the problems, and the most common modes of failure, it was necessary to undertake a general review of the range of pipe materials and jointing techniques currently being utilised for water distribution throughout the USA, UK and Europe. An appreciation of the numbers of joints by material types was also gained by considering the joint populations within these locations. With the range of joint types identified, a review was undertaken of the potential joint failure mechanisms, and these were related to the factors which may influence the condition of the joint. These general findings were supported by the results from a survey undertaken with various utility companies to obtain a picture of the range of pipe materials and joints within their distribution networks, and to obtain an appreciation of the problems they face with pipe jointing, testing and condition assessment. The methods of repair and rehabilitation employed by these utility companies were also considered.

Following identification of the problem joint types, consideration was given to some of the major factors that can influence their sealing characteristics, the potential failure mechanisms and how these could occur in practice. A number of approaches were taken including examination of field joints, sealing performance capability under load, finite element analysis modelling and subjecting several joints to mechanical loading and analysing the failure mechanism. This work helped to identify parameters that may lead to joint failures, to assist with defining the sensitivity and resolution required from any potential measuring devices associated with joint condition assessment.

A comprehensive literature and manufacturer search was undertaken. This covered current, emerging and potential technologies to detect, condition assess, repair and rehabilitate pipe joints. A more detailed assessment of selected technologies was then performed using a combination of controlled laboratory tests to establish the sensor performance, and field tests to understand the practicalities of utilising the various techniques in real pipe situations. The research approach acknowledged that location, condition assessment and repair of joints in pipelines is likely in practice to be achieved using a range of different techniques dependent upon pipe material, diameter and the particular operating conditions. With this range of potential techniques available, guidance for the selection of the most appropriate technique for the particular field operating conditions was developed in the form of a decision support tool.

RESULTS

The properties and characteristics of the most common pipeline materials and joint types used in water distribution networks were summarised and divided into three generic types:

- Metallic (grey and ductile cast iron, steel)
- Mineral based (asbestos cement and concrete)
- Plastic (PE, PVC and GRP)

Generally, a large proportion of networks still consist of cast iron pipelines with push-fit lead/yarn joints and the proportion of other materials is diameter and utility related. For the most common joint types, tables were developed to summarise the potential failure modes and causes. A utility survey questionnaire was developed to assist in developing a better understanding of joint failure problems faced by the industry. The general findings from the Utility Survey supported the initial search results with cast and ductile iron pipes dominant. Major observations were that problems were encountered with gasket/seal failure through degradation and corrosion of bolts and backing rings. Repairs generally consisted of open trenching and performing cut-out and replacement of the joint. Visual assessment of the condition of excavated joints is undertaken, with active leak detection and pressure/flow surveys.

For consistency, it was decided that joint condition would be ranked as a level or class of degradation, and the system adopted is described throughout the report. For each of the materials and joint types identified, all known degradation and failure modes were detailed and summarised in a series of tables. Understanding joint performance was summarised by a flow chart of parameters influencing joint condition. Analysis of field joints has shown that even when no yarn is present, the joint may still remain sealed due to the presence of the lead. Monitoring changes in joint gap and angular displacement could provide a useful indication of joint condition.

A rig was developed to try and gain a better understanding of gasket performance as a sealing medium for joints. The performance of several different gasket types was assessed, by varying the pipe cleanliness and the sealing forces. Also in order to better understand the limit state conditions of socket and spigot joints with elastomeric seals, a test was undertaken to record the strain induced in the pipe relative to applied bending loads and angular deflection.

Analysis of exhumed lead/yarn field joints subjected to mechanical loading to failure showed that generally, the first signs of leakage occurred between the pipe spigot and the lead within the joint. Parameters associated with the applied loading regime and the actual joint movement were also identified.

Finite element analysis modelling of push-fit lead/yarn joints was very complex, and the limitations of the modelling package used were evident. More accurate input data is required from laboratory testing and benchmarking to build up confidence in the modelling output.

In the technology survey, each of the potential technologies identified was reviewed for its suitability to provide the required information, and a summary documented. In some cases, little information was available on the use of the technology for this application and these technologies were subject to more detailed assessment using laboratory and field tests.

The potential techniques for the detection of joints were divided into several categories and some of the techniques evaluated in detail are shown below:

- Non-intrusive – surface (e.g. magnetic detection, conductivity assessment)
- Non-intrusive – on pipe (e.g. guided wave)
- Intrusive – in-pipe (e.g. caliper pig, hall effect sensors)

The potential techniques for condition assessment of joints were divided into the categories of non-detectable and detectable leakage, and some of the techniques evaluated in detail are shown below:

- Non-detectable leakage – external (e.g. pipe surveying, ultrasonics, guided wave)
- Non detectable leakage – internal (e.g. CCTV/sonar, low flux magnetic sensors, in-pipe ultrasonic inspection of joints)
- Detectable leakage – external (e.g. magic carpet)
- Detectable leakage – internal (e.g. Sahara)

Also within the technology survey, a review was undertaken of potential techniques for the repair of joints and these were characterised as follows with a few examples of the techniques in each category.

- Externally applied repair systems (e.g. cut-out, clamps, collars, encapsulation)
- Internally applied repair systems (e.g. resin sealants, mechanical sealing, robotics)
- Pipeline renovation systems (e.g. pipe segments, cured in place pipe, resin lining and PE lining)

Use of the techniques for this application has been summarised and a method developed to assist with the evaluation of potential techniques. The evaluation of all of the techniques identified during the technology survey has been summarised in a series of tables and flow charts to assist with the selection of the most appropriate technique. In order that water utilities achieve a more practical benefit from the project, a methodology for the selection of the most appropriate joint location, condition assessment and rehabilitation techniques for their particular case has also been developed. This is accompanied by a tool in the form of a CD ROM which allows the selection of techniques based on their technical attributes.

CONCLUSIONS

The major findings are shown below and conclusions have also been drawn from each of the various sections within the report.

- Due to the variable nature of joint construction and as laid condition, the work undertaken has shown that it would be very difficult to provide measurable parameters and limits that would give a quantitative estimation of time to failure. The degree of leakage from a joint is currently the most reliable indicator of condition. However, the "acceptable" limits of leakage may still vary due to the operating constraints of the particular utility.
- The report suggests a number of parameters that could be measured or monitored to give an indication of joint condition, and the relative probability of future joint failure e.g. if a joint has deflected and become metal bound or spigot pull-out has reached unacceptable limits.
- Given the current availability of techniques for condition assessment of joints, it is considered that specific joint condition assessment may not be cost effective. However, if undertaken as part of a more general pipeline condition assessment programme, it may become more economically viable.

Joint Statistics

- There are a vast number of different joint types associated with the pipeline materials currently operating within water distribution networks.
- Based on installed pipe lengths and individual spool lengths per pipe material, an estimated joint population has been established. This population is in excess of 330

million joints in service in around one third of the potable water pipelines in the USA and Canada, along with all potable water pipelines in the UK and France.

- Data suggested that joint problems account for around 6% of all water distribution pipeline failures (excluding service connections) which require utility action.
- For all pipe diameters, due to the population of cast iron pipelines still in service, it is suggested that push-fit lead/yarn joints are still the most numerous.
- For larger pipe diameters (e.g. above 16 inch), data suggested that the concrete pressure pipe population, with its associated joints, is the most numerous in the USA.

Utility Survey

- It was evident that recording of joint failures (failure types and causes) has not been rigorously undertaken by water utilities in the past.
- The limited response to the utility survey means that although the data set is not statistically robust, it did provide some useful information:
  - CI and/or ductile iron were generally dominant for all diameters, with local variations in other material types.
  - For CI, DI and PVC, push-fit socket and spigot joints are the most common.
  - For Steel and PE pipe, butt welded joints are the most common.
  - CI showed the largest number of joint failures in a typical year, even for larger diameters.
  - The dominant type of joint failure for all mechanically jointed pipes was gasket/seal failure followed by bolt failure.
  - Socket failure of CI and PVC pipes was also significant.
  - The primary cause of joint failure for mechanically jointed pipes was gasket degradation.
  - The primary cause of joint failure for mechanically jointed non-metallic pipes was ground movement.
  - Joint failures are mostly repaired by cut-out and replacement.

Understanding Joint Behaviour

- For the many different joint types in service, there are many common parameters, which contribute to joint degradation.
- An assessment of joint degradation can be undertaken by a ranking system based on leak classifications ranging from a non-leaking joint to a constant stream of water emanating from a joint.
- From the samples studied, iron socket and spigot joints are capable of performing satisfactorily with a wide variety of lead/yarn content, joint gap and angular displacement.
  - It seems impractical to develop techniques to measure the above parameters as they provide little information on condition or projected life before leakage occurs.
  - Monitoring changes in angular deflection or axial pull-out may provide an indication of joints that are most at risk of failure. This could be of particular use following seismic activity.

- The work undertaken on flange joints suggests that good gasket performance depends upon appropriately designed gasket, clean faces and the correct level of bolt torque to provide the necessary seating and sealing forces.
  - Once installed, there does not appear to be a readily accessible parameter that could be measured to provide an indication of bolted flange joint condition, other than that the rubber will degrade with time however:
    - on full faced flanges, an external ultrasonic technique may have the potential to assess the overall soundness of the joint.
    - access to components such as bolts can also provide useful information.
- From the tests performed on push-fit DI joints tests with elastomeric seals:
  - Small diameter pipes are more likely to fail in pure bending along the pipe barrel rather than crushing or bending failure at the joint.
  - Larger diameter pipes are more likely to fail in crushing adjacent to the joint rather than in pure bending.
  - The tendency for the joint area to re-round as a result of angular displacement from increased load, therefore maintaining good seal contact suggests that angular deflection would not necessarily lead to loss of sealing performance.
- In mechanical testing of CI lead/yarn joints:
  - The first signs of leakage were generally between the spigot and the lead. If disbondment or the presence of water between the lead and the spigot could be detected, this would be a potential method of condition assessment.
  - In stepped tensile (axial) loading to failure, the lead will pull out of the socket (if an external raised step is present on the end of the spigot) or the spigot will pull out of the lead if no step is present.
- Potential measurement parameters indicating the onset of failure due to axial pull out: spigot withdrawal, lead withdrawal, disbonding of interface between lead and iron, leakage.
  - In cyclic tensile (axial) testing, leakage was generally worse during the compression part of the cycle.
- Potential measurement parameters indicating the onset of failure due to cyclic axial displacement: disbonding of interface between lead and iron, leakage.
  - In the bend tests, the observed leakage path was at 90 degrees to the plane of bending; i.e. at the expected points of rotation of the spigot in the socket.
- The location of weeps on an exposed joint may give an indication of the plane of bending and therefore be useful when assessing the cause of failure.
- Potential measurement parameters indicating the onset of failure due to bending: variation in joint gap around circumference, angular displacement, spigot and socket becoming metal bound, disbonding of the interface between lead and iron, leakage.

Finite Element Analysis Modelling

- Modelling failure modes of cast iron lead/yarn joints proved to be very complex and the limitations of the NASTRAN modelling package used were evident.
- The modelling results suggested that for cast iron lead/yarn joints:
  - Stress concentrations are dependent upon the loading direction, lead geometry, pipe fillet radii and the contact properties such as friction/adhesion.

    - The peak stresses are developed in the lead until a joint becomes metal bound, after which the contact area distributes high stresses between socket and spigot.
    - Material properties are not generally the dominant factors in joint behaviour. Interfacial characteristics such as adhesion appear to be more important.
- To build up confidence in the FEA modelling output, more accurate input data is required from laboratory testing and benchmarking.

Technology Survey

- A number of techniques were identified from the technology survey for the detection of joints in water distribution mains. These were categorised as surface applied techniques, techniques that require intimate contact with the pipe and techniques that are intrusive (surface, on-pipe, and in-pipe respectively).
    - Of these techniques, only CCTV is currently regularly used within the water industry, although there are a number of other techniques used in other utility industries (or under development), which offer potential in this application.
- Detection of joints does not always imply location of the joint. For example, if the objective was to detect joints so that external repair clamps could be fitted, in-pipe joint detection techniques would only be valid if they could provide an accurate above ground location so that the joints could be excavated and repaired.
- In condition assessment of joints, the level of deterioration could be assessed for certain parameters e.g. rubber hardness, if the original and current values could be compared or monitored periodically.
- The use of various ultrasonic techniques offers potential to obtain useful information on the condition of joints.
- There are few dedicated systems currently being used within the water industry to externally examine the integrity of non-leaking metallic joints.
- There are a number of technologies used in other utility industries (or under development), which offer potential in this application by measuring particular joint characteristics such as angular deflection.
    - There are systems to externally examine the integrity of non-leaking PE butt fusion and electrofusion joints starting to be used within the water industry.
- There are few dedicated systems currently being used within the water industry to internally examine the integrity of non-leaking metallic joints other than techniques such as CCTV inspection and pressure testing.
    - There are a number of technologies used in other utility industries (or under development), which offer potential in this application by measuring angular deflection, misalignment and axial separation.
- There are a number of leak detection techniques available which would be appropriate for joint condition assessment if it could be confirmed that the leak was from a joint and not the pipe barrel.
- If the condition of the joint can be assessed, it would be possible to select the most appropriate repair/renovation technique according to condition
- There are a number of techniques already available within the water industry for the external repair of individual pipe joints, and there are several others used by other utility industries, which offer potential in this application.

- There are a number of internally applied systems for the repair and renovation of individual joints within both potable water and wastewater pipelines.
  - There are a number of resin and mechanically applied systems used in other utility industries (or under development), which offer potential in this application.
- Several of the techniques considered are generally for pipeline rehabilitation rather than individual joint repair. However, in situations where there are a number of leaking joints, they may be the most appropriate option, especially for smaller diameter pipelines.
- A weighting system can be utilised by utilities as an aid to evaluating repair and rehabilitation techniques.

Decision Support Tool

- To assist water companies in obtaining the most appropriate technical, practical and economic solutions, a logical approach to the selection of techniques for joint detection, condition assessment and repair is necessary.
  - A decision support methodology provides a source of reference and a logical approach to technique selection.
  - Although the methodology stands alone, a supplementary CD is provided to assist water utilities with the selection of techniques.

RECOMMENDATIONS

- This report has addressed the AwwaRF project objectives through consideration of the technical performance of current and potential systems for joint detection, condition assessment and repair. However, it is recommended that the economics of joint condition assessment are investigated in more detail to determine the potential utility market and therefore the viability of further developing techniques specifically for this application.
- Areas prone to seismic activity could be a potential market that should be investigated further, and may prove to be more cost beneficial.
- It is recommended that a study is undertaken to determine what level and detail of information on joint condition would be required by a utility asset manager to trigger remedial action.
- Utility companies should be actively encouraged to record data on joint failures and the data should be gathered annually and compiled into a comprehensive database.
- AwwaRF should produce a standard data set for recording joint failures and circulate this around member companies.
- Collection and examination of more field joints would enable a better understanding of the construction and failure modes of old joints.
- Destructive testing of old and new joints is recommended to gain a better understanding of how joints fail, the parameters to be measured or monitored, and the further development of techniques to enable this to be achieved.

- Further pressurised DI bend tests are recommended to look at sealing and failure modes combined (in an appropriate rig) e.g. to confirm that failure occurs outside the joint area in bending.
- To undertake more extensive modelling, a software package more applicable to this application will be required.
- A comprehensive programme of work should be undertaken to gain a better appreciation of the required input parameters for FEA modelling, and further joint types should be considered e.g. PCCP.
- Further consideration should be given to some of the techniques for joint location and condition assessment used in other industries, and a detailed assessment of the potential market and development required should be undertaken.
- Ultrasonic techniques should be investigated further as they have shown potential in determining the integrity of:
  - Rubber gaskets in flanged joints
  - Lead/pipe disbondment in CI lead/yarn joints
  - Fusion joints in plastic pipes
- Further consideration should be given to some of the techniques for joint repair and renovation used in other industries, and a detailed assessment of the potential market and development required should be undertaken.
- The decision support methodology should be used to assist with selection of techniques for joint detection, condition assessment and repair.
- Utility companies should be encouraged to provide feedback and additional local knowledge to enable the decision support tool to evolve.

CHAPTER 1
INTRODUCTION

INTRODUCTION TO THE PROJECT

The objective of this research project was to identify and document key problems associated with the failure of joints in water distribution pipelines and to investigate and report on the performance of existing and emerging techniques for the location, condition assessment and repair of these joints. Consideration has also been given as to how the information relating to testing and condition assessment can be best employed by the utility organisations. A decision support tool has been developed to assist with the selection of appropriate technologies.

BACKGROUND

Many water distribution companies have only limited knowledge of the condition of their underground water networks. In order to manage these networks efficiently, it is becoming increasingly important that water companies gain a better understanding of the current condition and performance of their buried assets. This increased knowledge will enable companies to better identify the risks they face, prioritise the needs for investment and determine the most appropriate remedial action. Buried pipes, valves and fittings can account for up to 70% of the total value of a Water Company asset. Failure can result in loss of supply to customers, major damage to public property and pose a severe local flooding risk and thus significant costs. To fully assess the condition of underground assets it will be necessary for water companies and their operators to undertake detailed programmes of inspection and diagnosis.

Typical pipe failures can be classified as either joint or barrel failure. These can result from poor pipe laying procedures, ground movement loading, environmental deterioration, and internal pressure surges. Pipe barrel failures have been the subject of considerable investigation over a number of years, but many existing distribution networks also have deteriorating joints that reduce the serviceability of the pipeline. Joints are recognised as an important and integral part of any water distribution network, and leaking joints are a common occurrence, but as these failures are not always catastrophic, they often remain undetected or are not acted upon. The detectable failures are observed as leaks or pipe breaks at joints, and may result in the deterioration of water quality, or in extreme cases, loss of supply to the customers. Mechanical joints are particularly problematic due to the complexity of sealing requirements, the need for some rotational capability and perhaps thrust restraint. The failure of joints can be attributed to a number of factors, including:

- Poor installation practice
- Differential settlement of adjoining pipes
- Stress cracking around the joint
- Ageing and degradation of sealing components
- Flange failure due to incorrect bolting procedure & subsequent stress cracking
- Bolt failure due to preferential corrosion

Examples of specific types of joint failure could include degradation of natural rubber gaskets and seals, breakage of contact surfaces with rigid lead joints, and corrosion and expansion of sulphur-based jointing compounds.

Many existing pipe networks are thought to have deteriorating joints that can reduce the serviceability of the pipeline (serviceability failures are observed as leaks or pipe breaks at joints, and may result in water quality deterioration or even loss of supply to customers). However, the condition of joints in water distribution pipelines has not generally been used as an indicator during the assessment of distribution network rehabilitation or replacement needs. Moreover, an assessment of the condition of the pipe barrel is normally used in deciding between replacement and refurbishment options, and in the selection of a suitable rehabilitation technology.

Current joint rehabilitation methods can be disruptive to both the environment and customers. The majority of available joint repair techniques are applied externally, requiring excavation to allow access to the joint. Existing internal repair systems are assembled manually inside the pipe, require man entry, and are therefore only applicable to pipelines of 20 inch (500mm) diameter and above.

OBJECTIVES

The objectives of the study have been to:

- Identify and document key problems associated with failure of joints in pipelines.
- Evaluate and report on the performance of existing techniques that could be utilised for inspection and condition-assessment of the wide variety of joints utilised throughout water distribution pipe systems.
- Recommend improvements and new capabilities for future joint condition-assessment technologies that can be applied to both new installations and older pipelines.
- Identify and investigate potential methods to repair joints without the expense and service disruption resulting from open-trench excavation.

RESEARCH APPROACH

The research approach acknowledged that location, condition assessment and rehabilitation of joints in pipelines was likely, in practice, to be achieved using a range of different techniques. The choice of technique is dependent upon pipe material and diameter, and the project generally considered the following common pipeline materials: Cast Iron (Grey and Ductile Cast Iron), Steel, Pre-stressed Concrete Cylinder Pipes (PCCP), and Plastic (PVC and Polyethylene). The study considered distribution pipelines in the diameter range of 2 inch (50mm) and greater.

The project also considered how the information relating to testing and condition assessment could be best used by the utility organisations. A decision support tool was developed that simplifies the selection of appropriate technologies, for joint detection, condition assessment and repair. More specifically, the research approach was to:

- Determine the types of joints which are currently giving rise to most problems, and which modes of failure are most common and/or costly

- Perform a survey of water and other related utilities to determine which methods are currently being used for joint location and condition assessment, and which repair techniques are currently being utilised.
- Utilise the results from the survey to assess the benefits, problems, and limitations of commonly-used joint location, condition assessment, and rehabilitation methods.
- Perform a literature and manufacturer search of current and emerging technologies used for location, condition assessment, repair, and rehabilitation of joints in all types of pipelines.
- Investigate and report on the performance of selected location, condition assessment, and rehabilitation technologies and recommend situations where each can be used appropriately. Pipe re-lining and methods that seal and strengthen the joints were also considered.
- Conduct utility-sponsored field-testing for selected location and condition assessment methods under a variety of realistic conditions.
- Identify any additional research that may be needed to address this issue.

WORK PROGRAME AND REPORT STRUCTURE

The work programme was carried out in five phases and these are described below along with the appropriate chapters within the report.

Phase 1 – Assessment of Requirements

In Chapter 2, a review of the range of pipe materials and associated jointing techniques being utilised for water distribution networks throughout the USA, UK and Europe was undertaken. An appreciation of the numbers and types of joints by material was also gained by considering the joint populations within these locations. For the purposes of the study, the pipeline materials considered were classified into the following groups:

- Metallic (cast iron, DCI and steel)
- Mineral based (concrete, PCCP, and asbestos cement)
- Plastics (PE/PVC/GRP)

Also within this phase, a study was undertaken into the potential joint failure mechanisms associated with the range of joint types identified during the jointing technique review. The potential failure mechanisms were also related to the factors, which may influence the condition of the joint.

Phase 2 – Utility Survey

The objective of the Utility Survey was to obtain a picture of the range of pipe materials and joints within distribution networks, and to obtain an appreciation of the problems faced with pipe jointing, the numbers of failures and any testing and condition assessment, undertaken by the various utility companies. The first task within this phase was to develop a questionnaire highlighting the pipeline materials and joints used as well as the numbers of joint failures and modes of failure. The methods of repair and rehabilitation employed by the various utility companies were also considered. Following the responses received, the data was analysed and presented in several standard categories for comparison. Comments on any major trends or differences in approaches were also highlighted. The results for this phase are summarised in Chapter 3 and Appendicies A to H.

Phase 3 – Joint Failure Analysis

Chapter 4 describes the work undertaken to improve understanding of the major factors that can influence joint sealing, and to consider potential failure mechanisms and how they can occur in practice. A range of tasks were undertaken within this phase to improve this understanding.

A number of joints removed from field pipelines were examined to identify any significant factors within the aged joint construction, which could have compromised joint integrity. A number of practical tests were also undertaken to monitor the sealing capability of joints under load. A number of joints were subjected to mechanical loading to failure, and the joints analysed to better understand the failure mechanism.

Finally, the most common joint type (lead/yarn push fit bell and spigot) was modelled using finite element analysis techniques, so that the sensitivity of each parameter could be assessed in relation to failure mode. This is also summarised in Chapter 4.

The objective of this phase of the work was to help to identify the parameters that may lead to joint failures, to assist with defining the sensitivity and resolution required from any potential joint condition assessment devices.

Phase 4 – Technology Survey, Performance Testing and Evaluation

The objectives of this phase were to undertake a comprehensive literature and manufacturer search of current, emerging and potential technologies for the detection, condition assessment, repair and rehabilitation of pipe joints and to perform a detailed assessment of selected technologies both in the laboratory and on site. Evaluation work undertaken is described, and the main findings discussed in each associated section.

Chapter 5 discusses the various potential techniques for joint detection and divides them into non-intrusive (mainly techniques for joint detection from above ground) and intrusive, where the tool has to be inserted into the pipe to provide the required output.

The techniques for condition assessment of joints were divided into two generic types. In Chapter 6, the review concentrates upon the range of potential techniques, which may be able to provide information on joint condition prior to detectable leakage occurring, and is subdivided into external and internally applied techniques. Chapter 7 considers some of the more novel leakage detection techniques because if the leak can be detected, and the joint position established at the same time, then leakage measurement could be used as an indication of joint condition.

In Chapter 8, the various pipe joint repair and rehabilitation techniques are considered and categorised into four generic types. Following a description of the technique, the discussion concentrates on their potential use within the water industry and the section also highlights a suggested method for selection of the most appropriate rehabilitation technique based upon pipe joint condition and operational requirements.

Phase 5 – Decision Support Tools and Methodologies

Chapter 9, provides a methodology that enables the findings from the preceding phases to be presented in a manner, that will enable the water utilities to achieve a more practical benefit from the project. The aim has been to provide a logical approach to selecting the most appropriate technique from the range of options for joint location, condition assessment and repair for the conditions determined in the field.

As well as the development and documentation of the methodology, a practical tool has also been developed in the form of a CD ROM. This guides users through the selection of the most appropriate joint location, condition assessment and rehabilitation techniques for their particular case, based on technical attributes

CHAPTER 2
ASSESSMENT OF REQUIREMENTS

REVIEW OF PIPE MATERIALS AND JOINTING TECHNIQUES

Introduction

The material and joint characteristics relating to potential failure mechanisms of pipes currently in use are described in the following sections. The tables provide an indication of the properties of each material and some of the most common jointing systems employed in terms of their anticipated operational performance.

Pit and Spun Cast Iron Pipes

Grey cast iron pipes were originally introduced into the market around 1900 and were at that time produced by casting molten iron in vertical sand moulds. Many water companies world-wide still have a very high proportion of their distribution networks constructed from these materials. The pipe walls are relatively thick, but not always of uniform thickness because of occasional misalignment of the central core mould. The main properties can be summarised as shown in Table 2.1.

Spun grey iron pipes were first introduced onto the market around 1930 and replaced cast or grey iron pipes. In this process, the molten cast iron was poured into cylindrical moulds, which are rotating at high speeds so that the pipe walls were formed by centrifugal force. The moulds used in the process would be either metal or sand lined, metal being used for the 3 inch size pipes and sand for all the other sizes up to 24 inch nominal internal diameter. By this process, a denser iron results and the pipes were of a more uniform wall thickness.

These spun iron pipes were in widespread use during the period from 1930 to 1960, and still comprise a large proportion of the distribution mains of water companies in many countries around the world.

Table 2.1
Grey iron/spun iron

Characteristics	Use
Advantages: • Low cost, but no longer used Disadvantages: • Structurally weak • Internal corrosion can lead to water quality problems • Age is not a good indicator of structural condition, older pit cast iron having thicker walls than more recent spun cast iron pipes	These materials would have been in widespread use before the introduction of ductile iron and the thermoplastic materials. The majority of the current water utilities pipelines world-wide are constructed of pit or spun cast iron materials.

Data adapted from WRc 1995.

The utility information gathered in Chapter 3 shows that a large proportion of the water pipelines operating around the world are constructed of pit cast and spun cast iron. It is known that the poor structural performance of these materials can lead to significant system failure and leakage levels. These failure rates can have a very significant influence on the asset value and operational costs of water supply systems.

The factors, which lead to general patterns of failures in iron mains, are either corrosion or high levels of induced strain, or most commonly a combination of both.

Induced strain results from both the maximum system operating pressure, and from externally applied loads transferred through the ground. Pipe systems are normally designed to accept the maximum system pressure, with a tested factor of safety of perhaps 1.5, and a corrosion allowance applied to the pipe wall. Therefore, assuming corrosion does not exceed expected levels, in a correctly designed system the maximum operating pressure should have little influence on the long-term integrity of the pipeline.

The predominant pipe stresses in pipe of less than 300 mm diameter are due to externally induced bending loads and these can lead to failure of smaller diameter iron pipes, even where there is negligible corrosion. However, in pipes of larger diameter, crushing failures predominate, resulting in longitudinal fractures.

Cast iron pipe, unless fully protected, is also susceptible to major corrosion problems and can suffer from general corrosion, localised pitting corrosion and fissure corrosion under applied external bending load.

Cast Iron Pipe Jointing Methods

Lead Yarn Joints

In water distribution networks the oldest known method of joining pipe sections is by the use of lead yarn joints. This type of joint first appeared during the middle of the 19th century, with the introduction of cast iron pipes into the utility market. This type of joint was phased out by the 1960`s. Over this period, little development occurred. The joint consisted of a section of tarred rope inserted into the socket, which was held in place by a wedge of lead, which was melted, poured and hammered into place behind the rope. In some cases, lead wool was used in place of molten lead.

Mechanical or Mechanically Assembled Joints

In this type of joint assembly, a compressive force is applied to an elastic element (gasket), which provides the seal. The compressive force is normally provided by an arrangement of bolts. There are a variety of clamping and sealing arrangements; the principal variations are described below.

Flanged Joints

For cast iron, the type of joint shown in Figure 2.1 is believed to be the oldest, often with polygonal flanges in various shapes, incorporating leather gaskets, which later gave way to the more modern sealing materials. The flanges evolved into the more common circular shape and raised faces on the flanges were introduced to intensify the clamping forces from the bolt arrangements. Although not now generally used for buried pipe work, due to the limited flexibility compared with other mechanical joint types, this type of joint is used for fittings and other pipeline components, providing accurate positioning, effective locking, and ease of removal for maintenance when required.

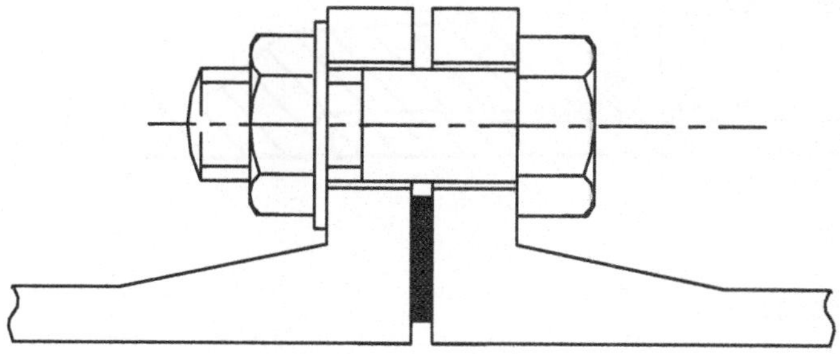

Source: Saint Gobain Pipelines.

Figure 2.1 Typical Bolted Flange joint

Sleeve or Coupling Joints

These joints are designed to solve the problem of connecting two cylindrical pipes of identical external diameter. The joints can be of different materials to the pipe and are commonly used for pipe repairs. This type of joint appeared around 1920, and is still in use today. An example is shown in Figure 2.2.

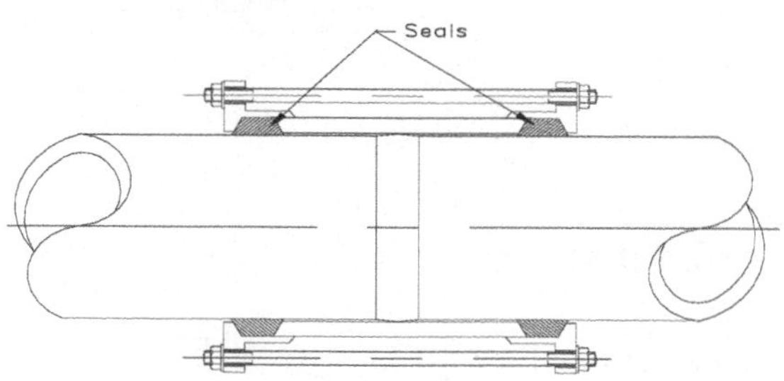

Source: Saint Gobain Pipelines.

Figure 2.2 Typical coupling joint

Joints with Gland Rings

This type of pipe joint employs the use of an elastomeric ring located in the socket of the pipe section. This is compressed by a gland ring reacting against bolts linked to the socket housing. Clamping force can also be applied by means of a screwed gland acting on a thread cut on the inside of the pipe socket assembly. Examples of this type of joint are illustrated in Figures 2.3 to 2.6.

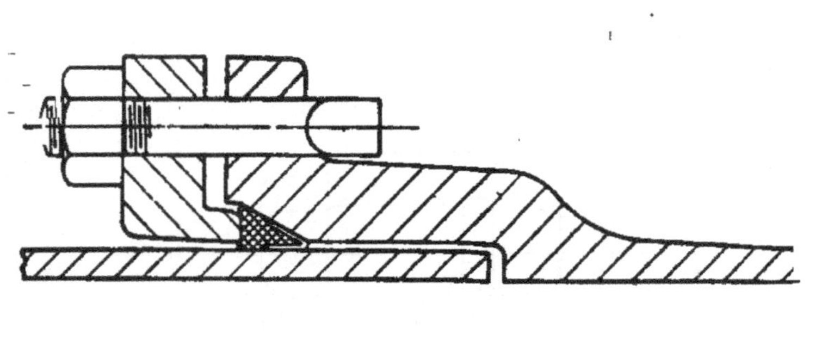

Source: Saint Gobain Pipelines.
Figure 2.3 Bolted Gland joint

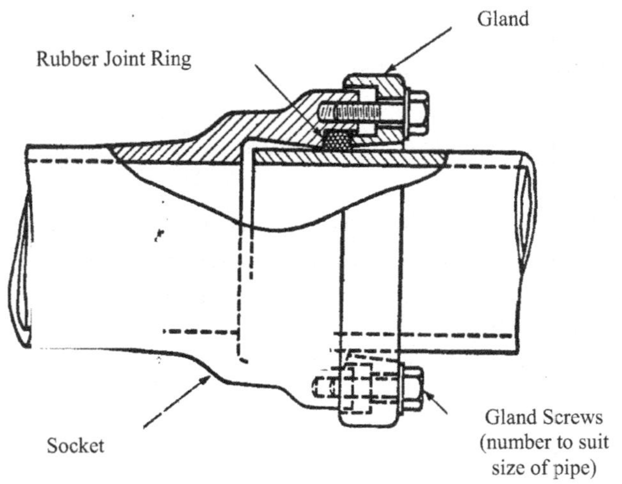

Source: Saint Gobain Pipelines.
Figure 2.4 Bolted Gland joint with set screws

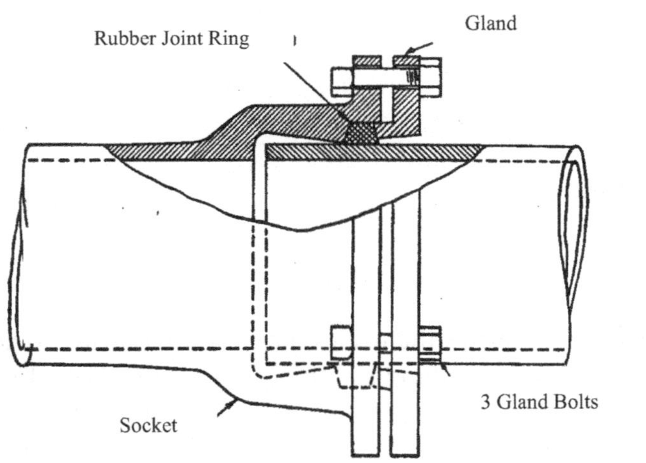

Source: Saint Gobain Pipelines.
Figure 2.5 Bolted Gland joint alternative joint gasket

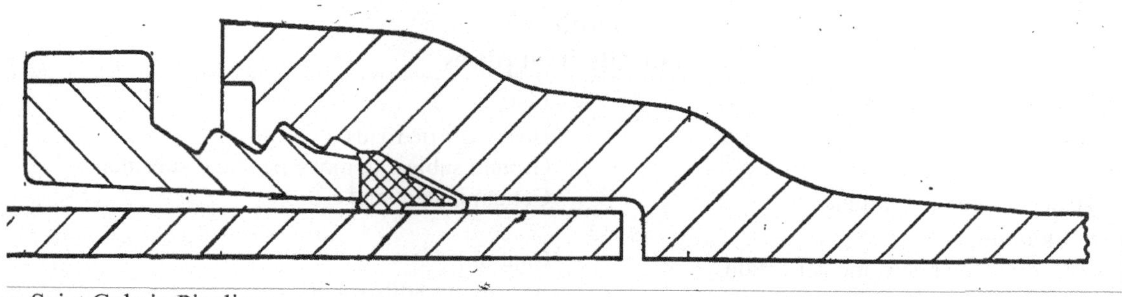

Source: Saint Gobain Pipelines.

Figure 2.6 Screwed Gland joint

Automatic Compression Joints with Elastomeric Joint Rings (Push-Fit Joints)

The joint is made by inserting a spigot into a socket containing a retained sealing ring. This type of joint appeared during the 1950`s and have largely replaced the alternative mechanical joints. Under service conditions, the increase in fluid pressure increases the contact pressure of the sealing ring against the spigot and socket surfaces, improving seating performance. An example is shown in Figure 2.7.

Unlike flanged joints or lead yarn joints, this type of joint can tolerate angular deflection after assembly. This provides a degree of resistance to the damage that can result from ground movement.

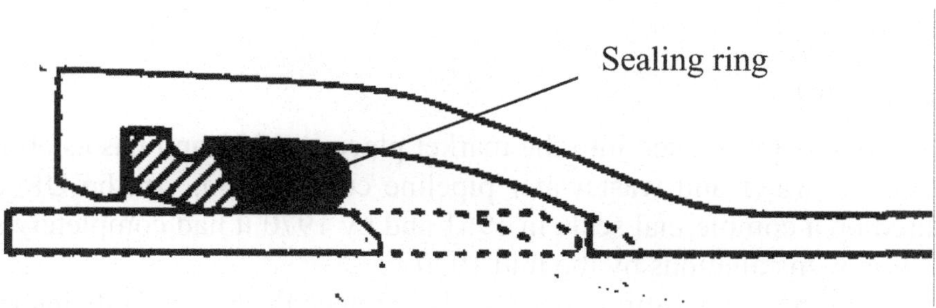

Source: Saint Gobain Pipelines.

Figure 2.7 Push-fit joint

Ductile Cast Iron Pipes

Ductile iron retains all of the attractive qualities of cast iron, such as machineabilty and corrosion resistance, but also provides additional strength, toughness and ductility. Although its chemical properties are similar to those of cast iron, the manufacture of ductile iron involves significant casting refinements, additional metallurgical processes, and superior quality control. The characteristics can be summarised as shown in Table 2.2.

Ductile iron also differs from cast iron in that its graphite form is spheroidal, or nodular, instead of the flake form found in cast iron. This change in graphite form is accomplished by adding an inoculate, usually magnesium, to molten iron of appropriate composition during manufacture.

Due to its spheroidal graphite form, ductile iron has approximately twice the strength of cast iron as determined by tensile, beam, ring bending, and bursting tests. Its impact strength and elongation are many times greater than cast iron.

Table 2.2
Ductile iron pipes

Characteristics	Use
Advantages: • Strength and toughness • Resistance to pressure fatigue • Ease of jointing • Flexible joints tolerate some deflection • End-load resistance joints available • Well established methods of repair • Impermeable to organic (gas+oil) contaminants • Easy location and leak detection • Imported bedding not required Disadvantages: • Requires protection against internal and external corrosion • Ground contaminants may cause failure of PE sleeving • Potential rise in pH when conveying soft waters • Corrosion protection systems at risk in certain soils • Corrosion protection systems susceptible to impact and accidental damage • Susceptible to attack by stray currents	Heavy traffic roads Ground subject to movement and subsidence Contaminated land Caution: Aggressive soils/waters Where there is a risk of stray (interference) currents Locations subject to high levels of third party interference (PE sleeved pipes)

Data adapted from WRc 1995.

Ductile iron was introduced into the market place in 1955 and has established itself as an Industry standard for water and wastewater pipeline construction. In the UK ductile iron was first manufactured on a commercial scale in 1961 and by 1970 it had completely superseded spun grey cast iron for new installations by the mid 1970.

Ductile iron pipes and fittings need to be protected against both internal and external corrosion, using recommended procedures. Cement mortar linings are commonly used for internal protection. External protection is generally achieved by the application of a zinc paint and outer coating or by means of a sleeve.

Although the pipes are robust, care of the protective coating is essential at all times. Ground conditions and third party interference are clearly significant sources of local damage and degradation leading to possible premature failure.

Ductile Iron Pipe Jointing Methods

Automatic Compression Joints with Elastomeric Joint Rings (Push-fit Joints)

This type of joint is very common where ductile pipe has been used, and is characterised by the radial compression of the elastomeric gasket by forcing the spigot end of one pipe, into the socket of the next pipe. The gasket is initially thicker than the annular space of its final housing inside the socket.

A modification of this type of joint has been developed, which is the push fit anchored joint which uses a modified gasket containing moulded in stainless steel teeth. Under pressure the joint tries to pull apart causing the teeth to grip the spigot preventing joint separation, as shown in Figure 2.8.

The anchor gaskets must only be used once, if the joint is dismantled for any reason a new gasket must be used to re-assemble the joint. Figure 2.8 illustrates an example of this jointing arrangement.

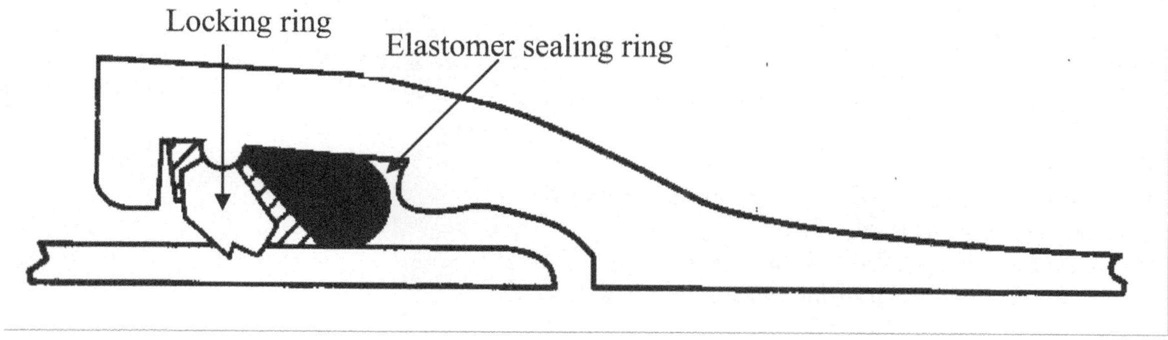

Source: Saint Gobain Pipelines.

Figure 2.8 Push-fit joint with locking ring

Mechanical or Mechanically Assembled Joints

As with the joints described for the pit and spun iron pipes, mechanical joint designs are also employed in ductile iron pipeline construction. For this type of joint, the sealing gasket is initially thinner then the annular space of its housing in the socket. The spigot of the adjacent pipe can be introduced without deformation of the gasket. The latter is then compressed axially by mechanical means, to provide and maintain in the long term, the required axial force. Design options are available which provide mechanical anchoring to push-fit joints.

Flanged joints are universally used with most types of pipe material, employing a gasket, which is clamped between the flanges and compressed, to affect a seal by an arrangement of nut and bolt assemblies.

Steel Pipes

Steel pipes and fittings must be appropriately protected against internal and external corrosion. Internal protection is achieved by the application of bitumen, cement mortar or concrete. External protection is achieved by bitumen application, sheathing or enamel, coal tar wrapping or polyethylene cladding, and care must be taken to minimise the possibility of damage to the protective coating. Damage to the linings can occur due to flexing if the pipe is not supported adequately, particularly during transportation. Thin walled pipe linings are more susceptible. Cathodic protection is recommended in conjunction with normal protection to safeguard against damage prior to installation and in service. Any breakdown of this protection is likely to lead to local degradation and possible failure. Characteristics of steel pipes can be summarised as shown in Table 2.3.

Table 2.3
Steel pipes

Characteristics	Use
Advantages: • Strength and toughness • Flexibility • Resistant to pressure fatigue • Ease of jointing • Impermeable to organic contaminants • Can be welded to form leak-free, end load resistant systems Disadvantages: • Requires protection against internal and external corrosion • Potential rise in pH with soft waters • Corrosion protection system susceptible to damage • CP systems require regular maintenance • Susceptible to attack by stray currents • Large diameter to wall thickness ratio requires good ground support • Thin walled pipe with mechanical fittings may suffer leakage	Contaminated land Heavy trafficked roads Areas with seismic activity Caution: Aggressive waters Low stiffness pipes with ground movement

Data adapted from WRc 1995.

Steel Pipe Jointing Methods

Figures 2.9 to 2.17 illustrate the large range of jointing methods, which can be employed when using steel as the pipeline material.

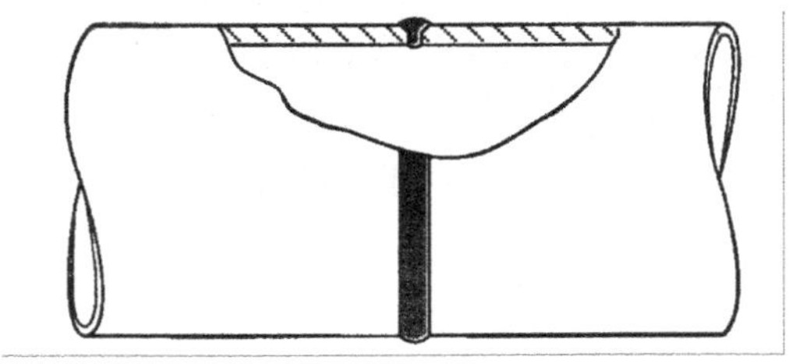

Source: WRc 1995.
Figure 2.9 Welded Butt joint

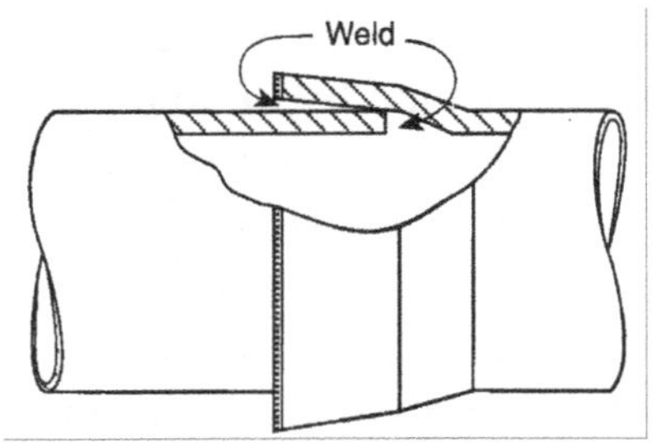

Source: WRc 1995.
Figure 2.10 Taper Sleeve welded joint

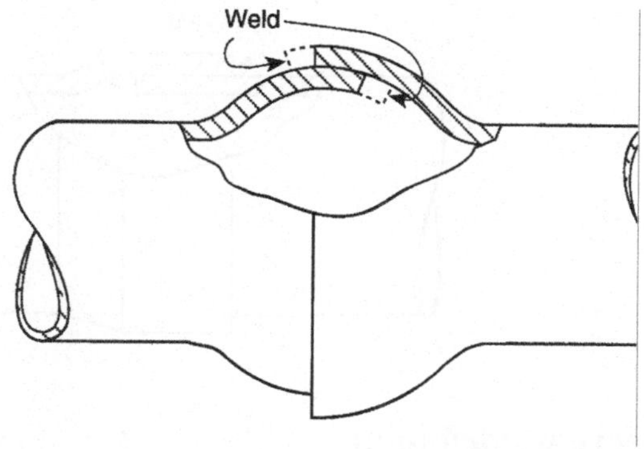

Source: WRc 1995.
Figure 2.11 Welded spherical joint

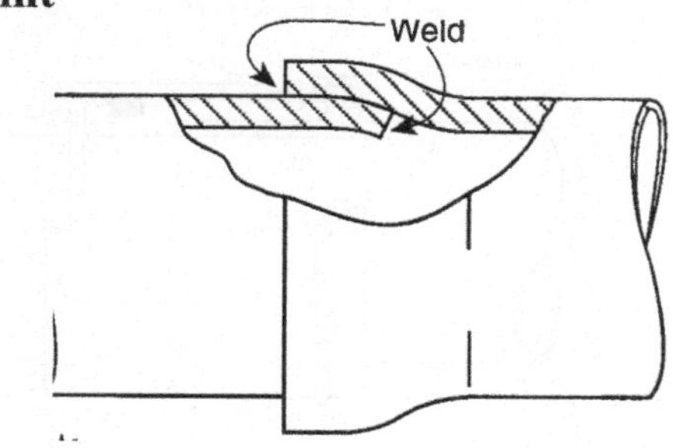

Source: WRc 1995.
Figure 2.12 Hemispherical welded joint

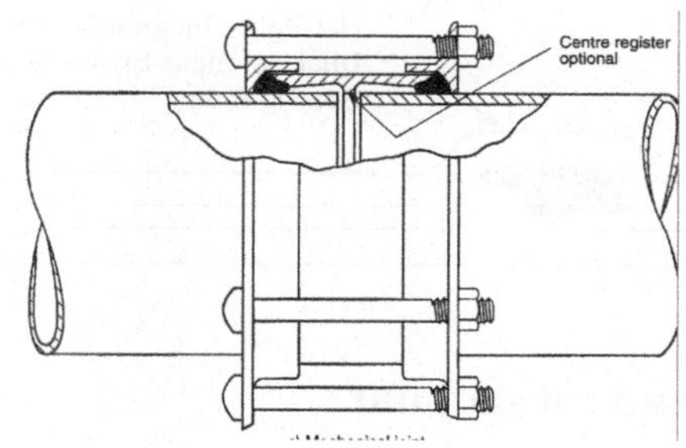

Source: WRc 1995.
Figure 2.13 Steel pipes with elastomerically sealed joint

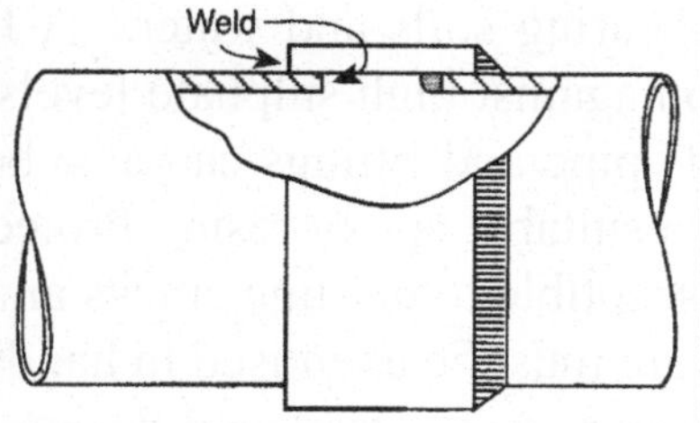

Source: WRc 1995.
Figure 2.14 Welded Collar sleeve joint

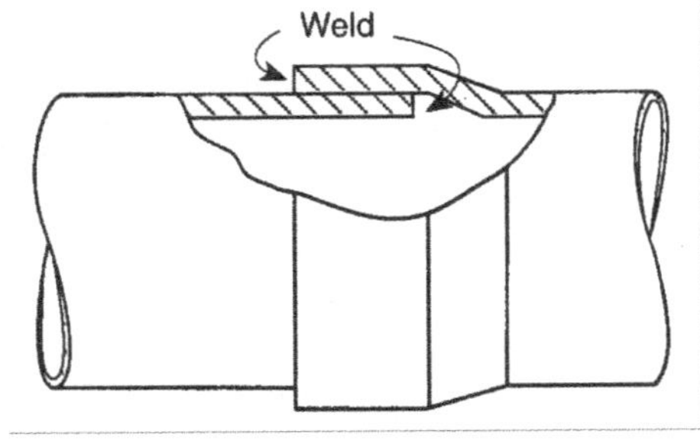

Source: WRc 1995.
Figure 2.15 Parallel Sleeve welded joint

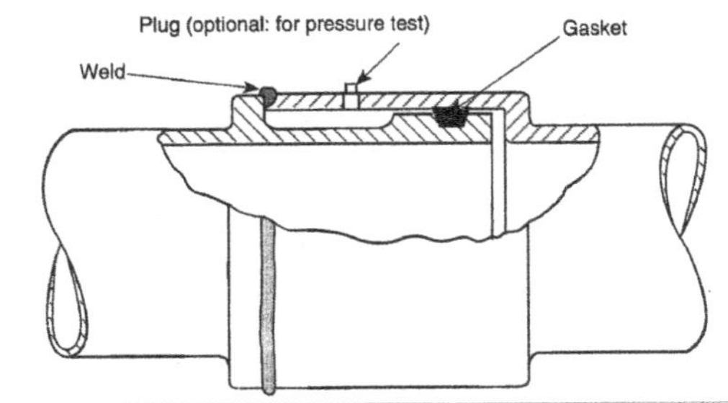

Source: WRc 1995.
Figure 2.16 Welded spigot and socket joint

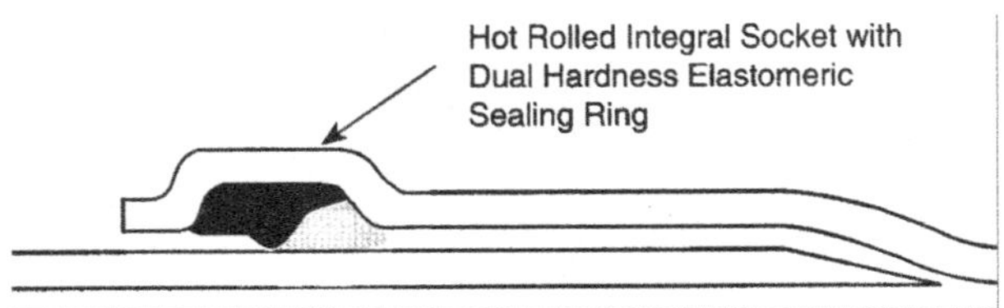

Source: WRc 1995.
Figure 2.17 Unrestrained spigot and socket joint

Asbestos Cement (AC) Pipes

Asbestos Cement pipes are reasonably resistant to corrosion but are susceptible to attack from acidic, soft or sulphate-bearing soils and water. A bitumen coating assists in controlling these problems. For protection against high sulphate levels, a sulphate resistant Portland cement coating may also be used. AC pipe and fittings can also be protected with a sleeving, or coated internally and externally with a suitable epoxy resin. Protected iron and steel fittings can be used on this pipe. AC pipes are susceptible to hairline cracks and crushed laminations, especially at or near pipe ends. Appropriate care must be exercised in handling, storage and transportation.

Asbestos cement (AC) pipe was introduced during the 1930's and in the period 1950 to 1960 was used as an alternative to cast iron pipe. The manufacture of AC pipe ceased in the UK in 1986, but sources of supply exist elsewhere in Europe to BS 486, 1981; this standard was originally published in 1933. In the US, many states banned its use during the 1990's. Characteristics of these pipes can be summarised as shown in Table 2.4. Asbestos cement pipe is produced from cement slurry containing approximately 2% by weight of chrysotile (white asbestos) fibres of selected size and Portland cement. The mixture, in the form of a continuous ribbon, is fed in layers onto the outer surface of a smooth rotating mandrel, where it is consolidated by pressure. Once the required thickness is achieved, the pipe is matured under water for several days before trimming to length and preparing for the joints.

Table 2.4
Asbestos cement pipe

Characteristics	Use
Advantages: • Strength and Rigidity • Withstand fluctuating pressure and surges • Imported bedding not required • Corrosion resistant to most soils and waters • Flexible joints tolerate some deflection • Ease of jointing	Heavy trafficked roads Ground subject to minor movement Contaminated land (organic)
Disadvantages: • Susceptible to impact and accidental damage • Vulnerable to chemical attack by certain soils and waters	Caution: Aggressive waters and soils Locations subject to high levels of third party interference Ground subject to subsidence/significant movement

Data adapted from WRc 1995.

AC Pipe Jointing Methods

A flexible jointing system is available, which allows some longitudinal and angular movement during installation and in service. In the UK, two push fit joints described as Widnes (Figure 2.18), and Comete (Figure 2.19), joints are available. Both use elastomeric sealing rings, which are of the stabilised `O` type with different cross sectional areas and shapes. A joint is completed by using a parallel collar with two sealing rings, one in each end of a collar.

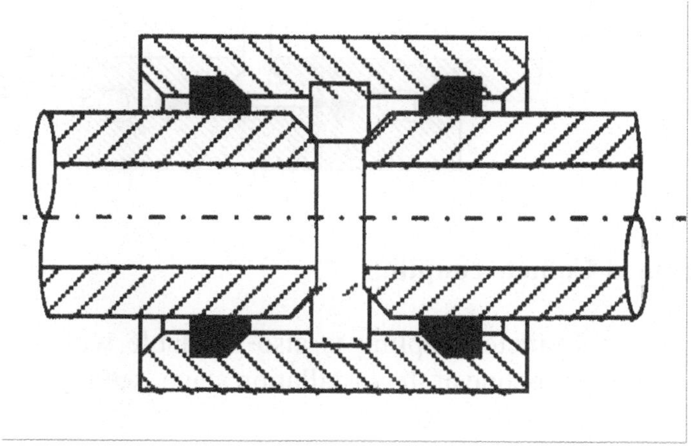

Source: WRc 1995.
Figure 2.18 Widnes joint

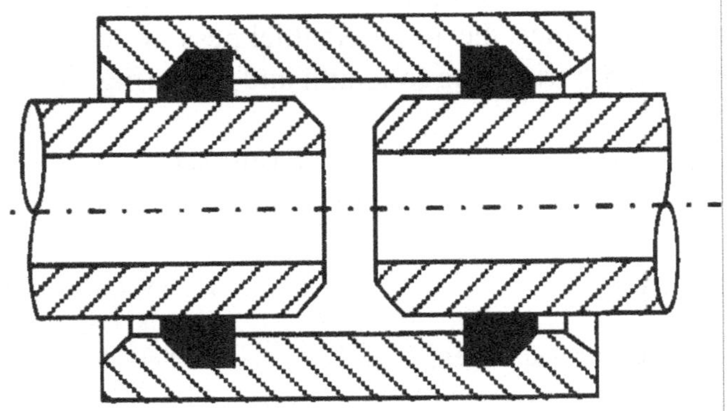

Source: WRc 1995.
Figure 2.19 Comete joint

The geometry of the collar and the pipes inserted within it will allow some degree of joint deflection to take place without leakage occurring. This is not an anchored joint although a version of the Comete coupling is available which incorporates an additional locking ring, which engages in recesses both in the pipe material, and in the wall of the joining collar, see Figure 2.20.

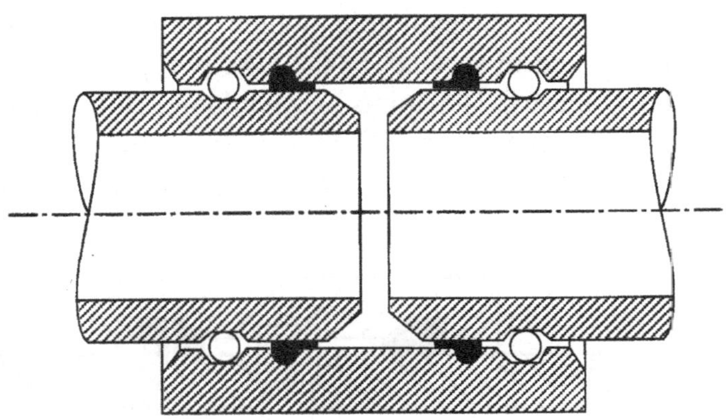

Source: WRc 1995.
Figure 2.20 Restrained Comete coupling

A version of the Comete coupling is also available with a built in offset to allow for changes in the route of the pipeline at the installation stage, see Figure 2.21.

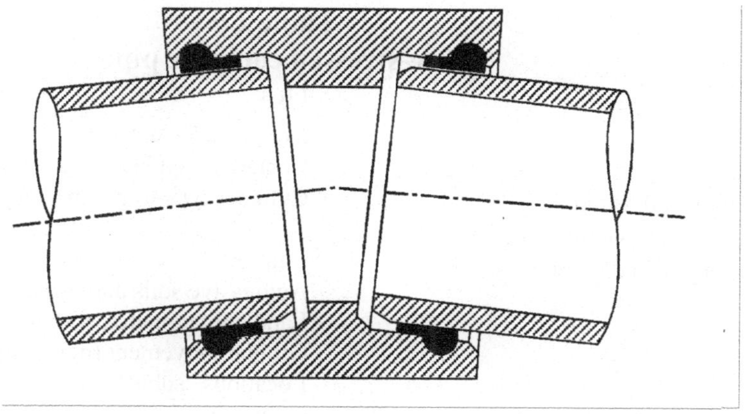

Source: WRc 1995.
Figure 2.21 Angled Comete coupling

Another joint is available which uses a cast iron or steel sleeve, compressed between two flanges and incorporating rubber-sealing rings, see Figure 2.22.

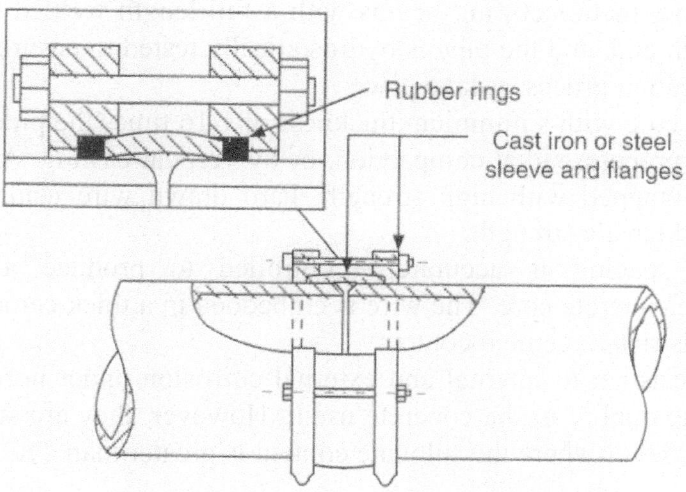

Source: WRc 1995.
Figure 2.22 Asbestos cement pipe joint with elastomeric seal

Pre-Stressed Concrete (PSC) Pipes

Pre-stressed concrete pipes were first developed in 1942 in the USA with Awwa standards C300 to C030 being developed in the late 1980's and early 1990's. Manufacture in the UK began in 1954, and the relevant British Standard was BS 4625, 1970. Several types of concrete pressure pipe (CPP) are manufactured and used for water distribution; these included pre stressed concrete cylinder pipe, reinforced concrete cylinder pipe and concrete bar wrapped cylinder pipe.

Because of their construction, some types of CPP are made for specific types of service whilst others are constructed in such a manner that a wide range of service conditions can be tolerated. Characteristics are shown in Table 2.5.

Table 2.5
Pre-stressed concrete pipes

Characteristics	Use
Advantages: • Corrosion resistance • High beam strength and rigidity • High degree of compaction not required • Impermeable to organic contaminants • Easy jointing system Disadvantages: • Heavy • Vulnerable to chemical attack from certain soils and waters • Joints are relatively inflexible	Heavy trafficked roads Contaminated land (organic) Cylinder type in ground subject to movement Caution: Aggressive soils and waters Land contaminated with neat solvents (non-cylinder) Ground movement (non cylinder) Locations subject to high third party activity (non cylinder)

Data adapted from WRc 1995.

Pre stressed concrete cylinder pipe (PCCP) has two general types of construction; a steel cylinder lined with a concrete core, or a steel cylinder embedded within a concrete core. In either type of construction, manufacturing begins with a full-length welded steel cylinder. Joint rings are attached to each end, and the pipe is hydrostatically tested to ensure there are no leaks before the concrete application processes take place.

A concrete core with a minimum thickness of 1/16 times the pipe diameter is placed either by the centrifugal process, radial compaction, or by vertical casting. After the core is cured, the pipe is helically wrapped with high strength, hard drawn wire using a stress of 75% of the minimum specified tensile strength.

The wire spacing is accurately controlled to produce a predetermined residual compression in the concrete core. The wire is embedded in a thick cement slurry and coated with dense mortar that is rich in cement content.

They are resistant to internal and external corrosion under normal conditions but this is dependent upon the quality of the concrete used. However, they are susceptible to attack where the pH is less than 5.0 or where the sulphate content is greater than 2%.

PSC Pipe Jointing Methods

The standard joint configuration for this type of pipe is essentially of the socket and spigot type sealed by means of an elastomeric gasket. The pipe spigot ring has a rectangular recess machined in it into which the ring gasket is located.

On assembly of the joint, the gasket is compressed as the spigot end of the pipe is pushed into the socket. The gasket diameter and the size of the socket and spigot ends are controlled to tight limits to ensure a reliable high-pressure seal is obtained on assembly.

Protective grout is applied to the joint after assembly in the field. Changes in direction can be achieved by the use of pre-cast connecting sections with the appropriate angles of deflection cast in at manufacture. Examples of the joint types are shown in Figures 2.23 and 2.24.

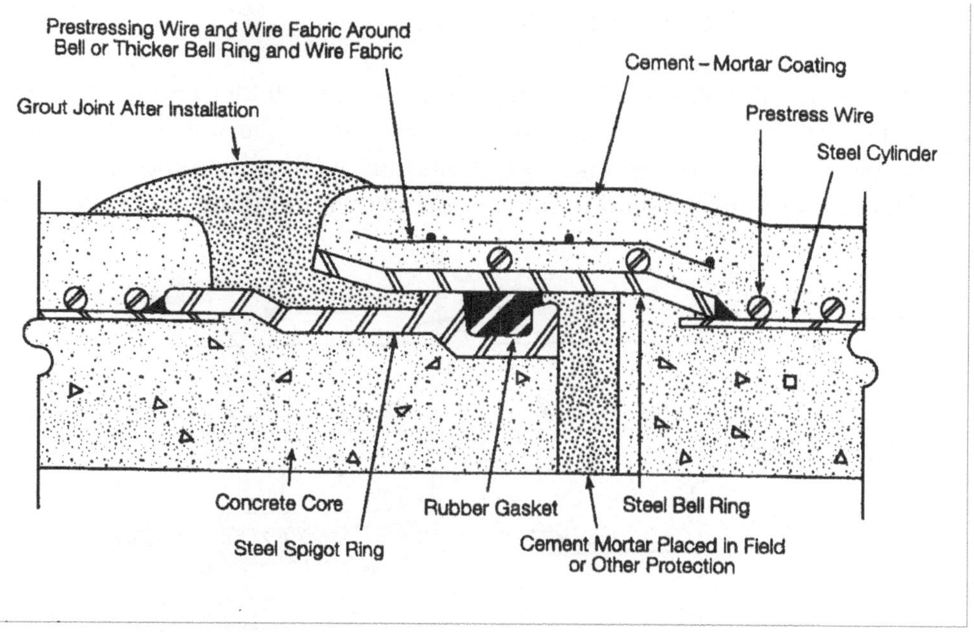

Source: Reprinted from Concrete Pressure Pipe (AWWA M9) by permission. Copyright © 1995, American Water Works Association.

Figure 2.23 Pre-stressed concrete Lined Cylinder pipe joint

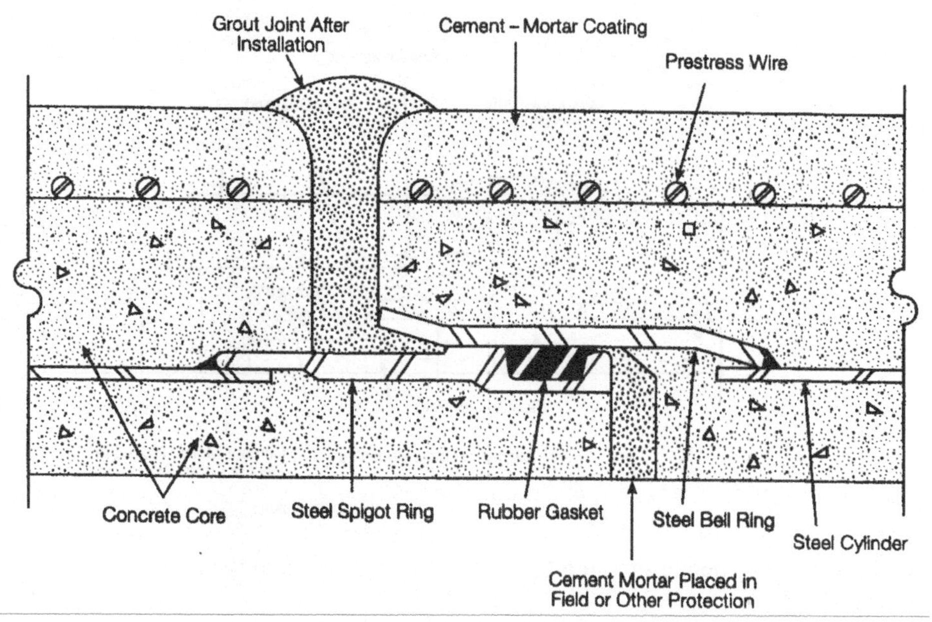

Source: Reprinted from Concrete Pressure Pipe (AWWA M9) by permission. Copyright © 1995, American Water Works Association.

Figure 2.24 Pre-stressed concrete Embedded Cylinder pipe joint

Reinforced Concrete Pipe Jointing Methods

Prior to the introduction of pre stressed concrete cylinder pipe in the 1940's most concrete pressure pipe was of reinforced concrete construction. The construction methods are similar except that the mild steel reinforcing is cast into the wall of the pipe instead of the pre stressing process. Examples of the standard joint types are shown in Figures 2.25 and 2.26.

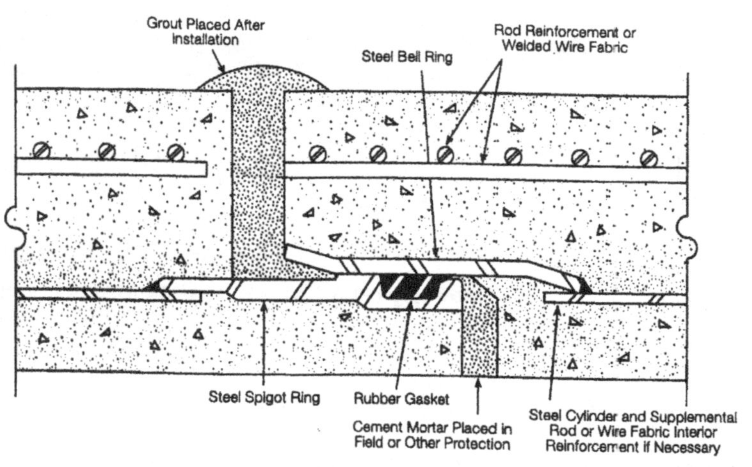

Source: Reprinted from Concrete Pressure Pipe (AWWA M9) by permission. Copyright © 1995, American Water Works Association.

Figure 2.25 Reinforced concrete cylinder pipe joint

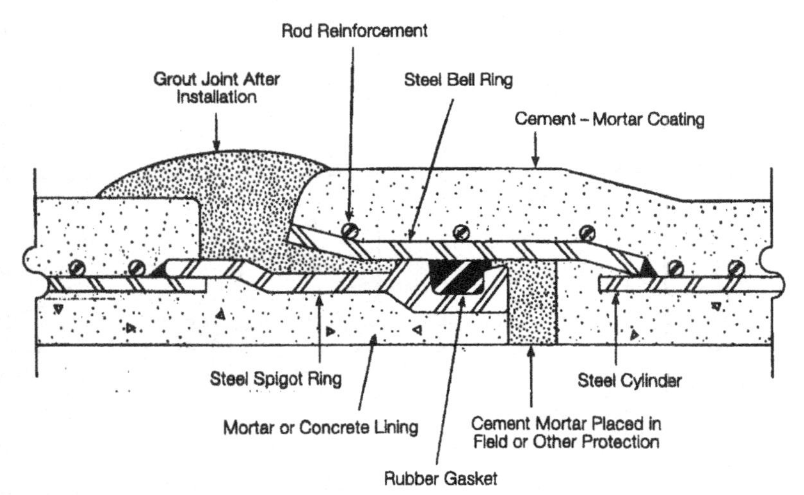

Source: Reprinted from Concrete Pressure Pipe (AWWA M9) by permission. Copyright © 1995, American Water Works Association.

Figure 2.26 Concrete Bar Wrapped Cylinder pipe joint

Under certain circumstances, as with other types of joints without end restraint, it is possible for this type of joint to pull apart and hence several designs of restrained joints are available. Details of the various types are presented in Figures 2.27 to 2.29.

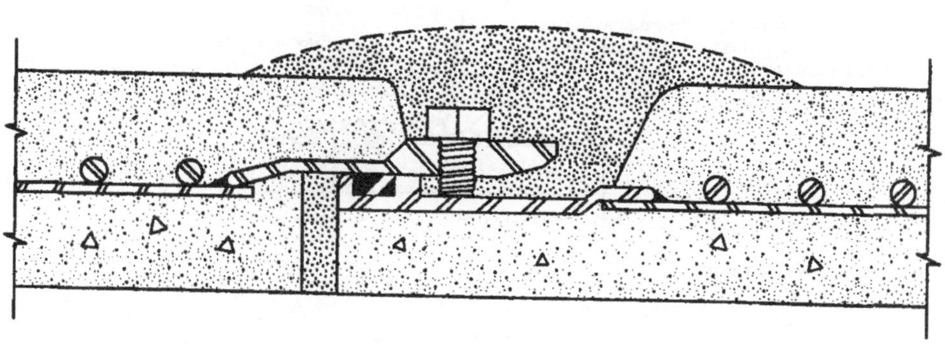

Source: Reprinted from Concrete Pressure Pipe (AWWA M9) by permission. Copyright © 1995, American Water Works Association.

Figure 2.27 Bell Bolt Harness

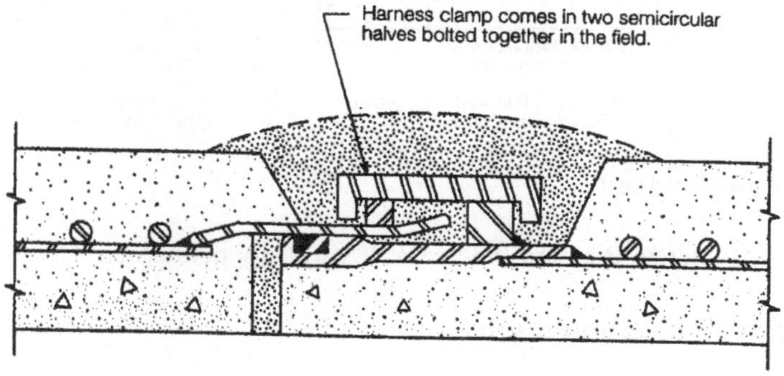

Source: Reprinted from Concrete Pressure Pipe (AWWA M9) by permission. Copyright © 1995, American Water Works Association.

Figure 2.28 Clamp Ring Harness

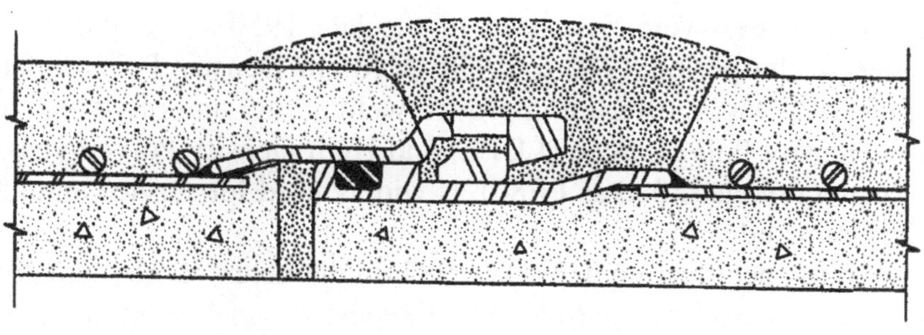

Source: Reprinted from Concrete Pressure Pipe (AWWA M9) by permission. Copyright © 1995, American Water Works Association.

Figure 2.29 Snap Ring Harness

As an alternative to the mechanically restrained options indicated above, welding of the steel cylinders can be carried out in a variety of ways as indicated in Figure 2.30.

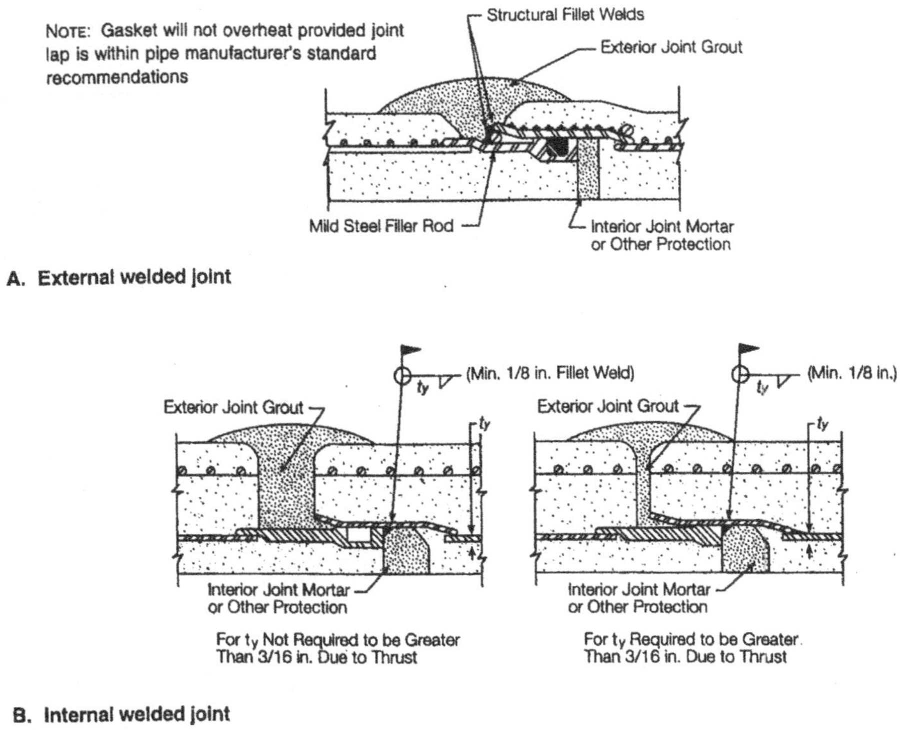

Source: Reprinted from Concrete Pressure Pipe (AWWA M9) by permission. Copyright © 1995, American Water Works Association.

Figure 2.30 Weld bead configurations

When it is evident that a joint in this type of pipe has developed a leak, then welding the steel cylinders together, after removal of the exterior grout, is an approved repair option.

Polyethylene (PE) Pipes

Polyethylene pipes were first used in the late 1950's and the more modern materials became available in the period 1980 to 1990. Medium Density Polyethylene (MDPE) became available in the early 1980's, offering improved weld ability and resistance to crack growth over the earlier Low Density Polyethylene (LDPE) and High Density Polyethylene (HDPE) materials. This material (generally classified as PE80), is widely used for service and distribution mains for pipe sizes of 250 mm and below.

PE80, to WIS specification 4-32-03, 1987, in diameters larger than 250 mm is down-rated to a lower operating pressure to guard against the potential risk of rapid crack propagation under extreme operating conditions, (thick wall pipes are at the greatest risk). Higher Performance Polyethylene (HPPE), classed as PE100 to WIS specification 4-32-13, 1993, was introduced in the early 1990's, providing a material with a higher strength rating, improved resistance to slow crack growth and improved resistance to rapid crack propagation.

The stress life characteristics of PE pipes are predicted by the use of a regression procedure. The regression plot shows the performance ability of the material under a specified load for a design life of 50 years. This method identifies a minimum required strength value (MRS); a safety factor is applied to the MRS to determine the design stress. If we consider a minimum permissible design safety factor for PE pipe of 1.25 for a 50-year design life by means of an example, a PE100, SDR11 (that is with a ratio of the pipe diameter to wall thickness of 11) pipe has a pressure rating of 16 bar. The permissible range of pressure fluctuation (or surge pressure) should not exceed 8 bar and the total pressure should not exceed 16 bar. Characteristics of the material can be summarised as shown in Table 2.6.

Table 2.6
Polyethylene materials

Characteristics	Use
Advantages:	
• Corrosion resistant	Most low and medium stress applications:
• Lightweight	Gravity mains
• Flexible	Light traffic roads
• Can be welded to form leak free end load resistant system	Ground subject to movement and subsidence
• Toughness	
• Absorbs impact loads, vibration and ground movement	
• Suitable for narrow trenching and trench less applications	Caution:
• Suitable for rehabilitation	Pumped mains with surge conditions
	Major carriageways
Disadvantages:	Contaminated land
• Susceptible to permeation/degradation by certain organic contaminants	
• Fusion jointing requires special equipment and training	
• Selected /imported bedding required	
• Reliant on stable support from ground to resist deformation	
• Susceptible to UV degradation	
• Risk of flotation	
• Mains location and leakage detection not established	
• Needs to be derated >250mm diameter if risk of long line fracture exists	

Data adapted from WRc 1995.

PE Pipe Jointing Methods

A fusion jointing process is normally utilised for PE pipes. Since the introduction of this pipe material, the techniques have been refined and developed. Essentially these processes involve melting the pipe material under controlled conditions and allowing the two elements of the joint to fuse together and make the joint. The joints can be either, fused socket joints, butt fused joints, or electro fusion joints. The fused socket jointing system would have been used with the earlier pipe assemblies, giving way to butt and electro fusion. The socket joint was completed by heating the inside of the PE socket and the outside of the pipe to be joined for a prescribed period and then pushing the pipe into the socket to finish the process.

With the butt fusion process, the end of the pipes to be joined are trimmed to ensure a full contact area and then heated by placing them in contact with an electrically heated platen for a prescribed time before pushing together under a controlled pressure. This process is now either fully automatic or semi automatic, to reduce the risk of poor joint quality, and to provide a record of each joint as it is completed. With this process, similar materials with the same wall thickness should only be used.

The electro fusion process is the latest joining technique and is fully automated to ensure that consistent joint quality is maintained. The electro fusion fitting has a heating coil built into it. When the pipes to be joined are placed in the fitting, it is energised for a prescribed time period during which the coupling and pipe sections are fused together. Correct procedures are shown in Figure 2.31.

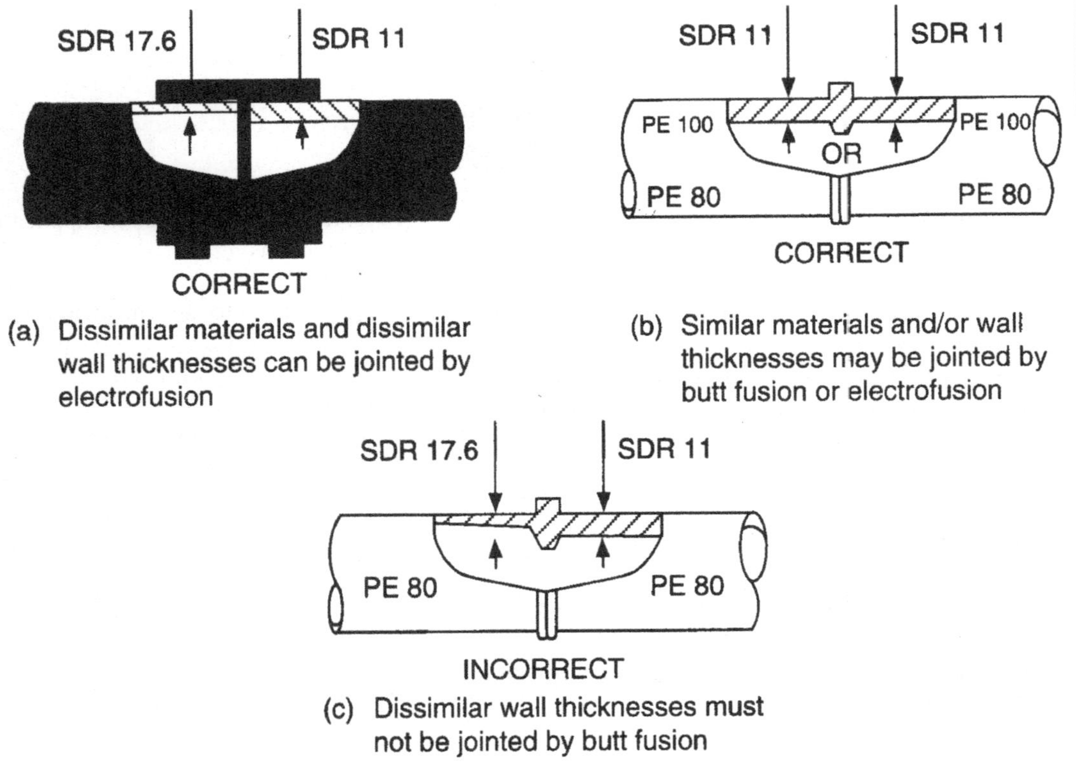

Source: WRc 1995.

Figure 2.31 Range of pipe joints for use with PE pipe (a) Electro fusion, (b) Butt Fusion (c) warning re Butt Fusion

A range of mechanical fittings can also be utilised for joining PE pipes to valves etc.via flange connections.

Polyvinyl Chloride (PVC) Pipes

Unplasticised polyvinyl chloride (PVC-U) was first introduced in the 1950's for water distribution. During the 1960's it came into widespread use and although the original material experienced some problems, the advantages of using PVC-U have ensured its widespread use to the present day. Three types of PVC pipes are now available:

- *PVC-U* Unplasticised PVC. The early unplasticised PVC's suffered from being somewhat brittle. Their tensile strength required that the pipes had thick walls to provide the necessary pressure capability. This reduced their economic competitiveness.
- *PVC-O* Molecular Orientated PVC. By orientating the molecules during the manufacturing process, pipe can be produced with thinner walls but equivalent toughness and pressure capability.

- *PVC-A* Modified PVC (PVC/PE alloy). By alloying PVC and PE, an alloy can be produced which provides thinner wall pipe with the necessary toughness and pressure capability.

PVC-U pipes manufactured in accordance with BS 3505, ASTM D2241 and Awwa C900, are the most commonly used for water distribution. Pipes to BS 3505 are available in three pressure ratings, 9bar, 12 bar and 15bar. General PVC-U pipe properties can be summarised as shown in Table 2.7.

Table 2.7
PVC materials

Characteristics	Use
Advantages:	
• Corrosion resistant	Most low stress applications;-
• Lightweight	Gravity mains
• Flexible joints tolerate some deflection	Light traffic roads
• High resistance to abrasion	
• Resistance to chemical attack	
• Ease of handling	Caution:
	Pumped mains with surge conditions
Disadvantages:	Major carriageways
• Susceptible to point loading and impact damage	Contaminated land
• Susceptible to failure during tapping	Ground subject to movement and subsidence
• Susceptible to permeation/degradation by certain organic contaminants	
• Certain grades susceptible to UV degradation	
• Risk of fracture in contaminated land	
• Joints need to be anchored	
• Selected /imported bedding required	
• Reliant on stable support from ground to resist deformation	
• Requires care in handling	
• Risk of flotation	
• Mains location and leakage detection not established	

Data adapted from WRc 1995.

PVC Pipe Jointing Methods

When the pipe material was first introduced, solvent cemented joints were the preferred jointing method. They gave way to simple O ring seals eventually leading to a variety of elastomeric lip seal type joints. The problem with these early joints was in being able to accommodate a wide range of tolerances for pipe spigot and socket, including pipe ovality and the tolerances associated with the sealing gasket itself.

Under certain circumstances, the sockets could expand, distort or deform leading to gasket movement and possible loss of sealing. In the mid-1970's, the "Rieber" jointing system was developed in Norway and eventually adopted world -wide.

The "Rieber" jointing system uses a gasket, reinforced with a steel wire, which is fitted into the socket end of the pipe and shrunk onto the spigot. The seal is factory fitted. An example is shown in Figure 2.32.

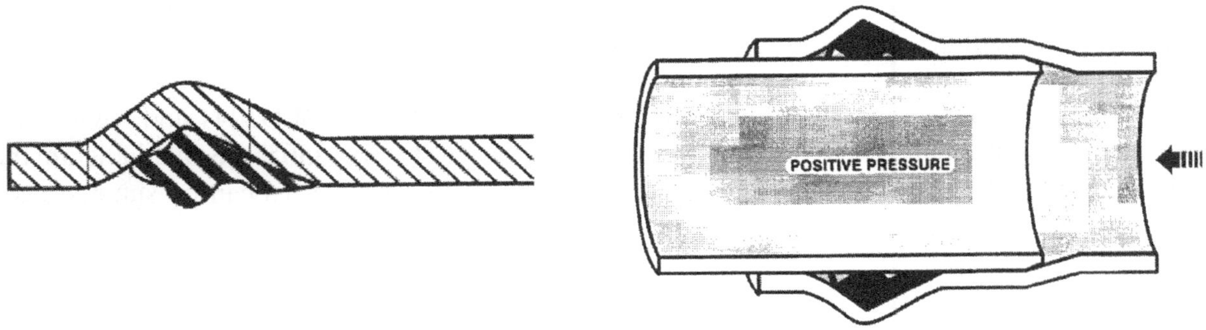

Source: Uponor Ltd.

Figure 2.32 Details of Rieber gasket and pipe joint assembly using Rieber gasket

Glass Reinforced Plastic (GRP) Pipes

Glass Reinforced Plastic (GRP) pipe is currently used in the UK water supply network, primarily for larger diameter trunk mains. Filament wound pipe became available late in the 1960's and centrifugally cast pipe became available in 1970. Most of the pipes used in the UK are centrifugally cast.

GRP pipes are not susceptible to general corrosion from naturally occurring soils but they are susceptible to strain corrosion in acidic conditions where the pH is less than 4.0. This should be accounted for at the design stage to keep pipe wall strains within safe limits. The use of thickened reinforced liners may also be used. Some degradation of GRP pipes can occur in contaminated soils, especially where certain organic products are present. Protected iron and steel fittings are used. The iron fittings have a cement mortar lining and outer PE sleeve.

GRP pipes are vulnerable to impact damage and need to be lifted with fabric straps or non-metallic ropes, and not be dropped, rolled or slid over rough hard ground. Visual inspection of the internal and external surfaces prior to installation allows identification of signs of partially cured material, uneven surface, air bubbles or blisters. Jointing surfaces must be undamaged.

High local strains can be generated during the installation process, and although the flexible joints accommodate some longitudinal or angular movement, the pipe must be well supported with good compaction of the reinstatement. Little information is available on the behaviour of GRP pipes under cyclic loads or surge conditions. Pipe characteristics can be summarised as shown in Table 2.8.

Table 2.8
Glass reinforced plastic pipe

Characteristics	Use
Advantages: • Corrosion resistant • Lightweight • Flexible joints tolerate some deflection Disadvantages: • Susceptible to impact and accidental damage • High standard of compaction • Strain corrosion attack in acidic conditions • Susceptible to permeation by contaminants • Little knowledge of performance under cyclic loading conditions	Minor carriageways Caution: Ground subject to subsidence or significant movement Contaminated ground Locations with high third party activity

Data adapted from WRc 1995.

GRP Pipe Jointing Methods

Many designs of pipe joints are available for GRP pipe materials and some are described in the following sections.

Sleeve Collar Joint

The standard GRP pipe system uses a spigot and socket coupling arrangement, which uses a GRP collar fitted with a synthetic rubber membrane liner as an integral part. They are fitted by pushing the collar onto the pipe to be joined, to form a socket into which the next section of pipe can then be pushed.

Harness Joint

The Harness joint consists of a metal harness with tie rods and nuts as appropriate. The joints are used where end thrust restraint is required.

Flanged Joint

Where connection to flanged units (such as valves and other pipeline hardware) is required in the pipeline, flat faced or stub flanges are available. These joints can also be used in pipelines to resist longitudinal loads, non-standard bolt patterns can be used.

Butt and Strap Joint

A butt and strap joint is a connection fabricated using resin and glass reinforcement, to form a continuous or longer length of pipe.

Mechanical Couplings

Standard mechanical couplings (similar to Figure 2.2), consisting of rubber seals and tightening systems can be used for tie-ins and connections to other pipe materials.

Flange Adapter

A flanged adapter is a mechanical coupling to enable a plain-ended pipe to be connected to a flanged pipe or fitting.

Ductile Iron Tie-in Joints

The outside diameter of GRP pipes up to and including 600mm nominal diameter are compatible with standard ductile iron sockets to allow connection of DCI and GRP.

JOINT STATISTICS

The joint population has been estimated from predictions of installed trunk and distribution mains within the UK, USA and countries in Europe for which data has been obtained.

UK Pipeline Population

The information for the pipeline population in the UK has been split into distribution (Table 2.9), pipes below 300mm diameter and trunk mains populations (Table 2.10) above 300mm.

Table 2.9
Length of distribution main (km) by material type

Material type	Length of main (km)		
	1983/84	1992	% of total (1992)
Cast and Spun Iron	222,750	191,250	64.3%
Ductile Iron	23,760	34,021	11.5%
PVC u	23,760	35,240	11.9%
MDPE	0	9,852	3.3%
HPPE	0	875	0.3%
Mo PVC	0	70	0.02%
AC	14,850	13,305	4.5%
Others	11,880	12,387	4.2%
Total	297,000	297,000	

Source: WRc 1987.

Table 2.10
Length of trunk main (km) by material type

Material Type	Length of main (km)		
	1983/84	1992	% of total (1992)
Cast and Spun Iron	16,500	13,900	63.2%
Ductile Iron	3,700	5,224	23.8%
PVC u	425	505	2.3%
MDPE	0	630	2.9%
HPPE	0	73	0.3%
Mo PVC	0	4	0.02%
AC	625	668	3.0%
Others	750	996	4.5%
Total	22,000	22,000	

Source: WRc 1987

The results of this survey indicate that although major changes in material usage have taken place over the last 20 years, a vast amount of cast and spun iron pipeline is still in place in the UK. This is considered to be a worldwide trend, with some regional variations due to system age and availability of materials.

USA Pipeline Population

In 1996, AWWA carried out a survey of its members with the aim of developing a database of water company statistics. 898 utilities (28% of the 3,200 surveyed) responded. Information on water mains material usage in the USA has been obtained exclusively from the AWWA "Water:\Stats" database (Table 2.11). Although this data represents less than a third of the network in the US and unfortunately, cannot be broken down by pipe diameter, it does give some indication of the proportions of different materials within the US systems.

Table 2.11
Type of mains in USA by material (all diameters)

Material	Amount of mains (miles)	Percentage of total
Asbestos Cement (AC)	56,360	15.0
Cast Iron (CI)	155,038	41.3
Ductile Iron (DI)	81,119	21.6
Pre-Stressed Concrete (PSC)	15,921	4.2
Glass Reinforced Plastic (GRP)	422	0.1
Polyethylene (PE)	1,318	0.4
Polyvinyl Chloride (PVC)	42,125	11.2
Steel	16,415	4.4
Other/unknown materials	6,719	1.8
Total length of mains	375,437	

Source: Reprinted from WaterStats, by permission. Copyright © 1996, American Water Works Association

It is clear from this data, that grey cast and ductile cast iron are the most significant materials in use, accounting for some 63% of the total mains surveyed. Although a great deal of this usage is likely to be in the smaller diameter sizes on distribution networks, there will be a significant amount laid in the larger sizes.

The next most popular material class is the mineral-based pipe, with asbestos cement and pre-stressed concrete making up over 19% of the total. In the larger diameter ranges, PSC is widely used. In the USA, it is manufactured in a range of diameters from 410mm (16in.) to 6.40m (21ft). Estimates for the amount of PSC pipe laid in the USA since 1945, reveal a figure of somewhere between 10,000 and 20,000 miles. In pipe diameters larger than 48-in., it is the most widely used high-pressure water pipe product in the country.

PVC, which accounts for 11% of pipes in the ground is much more likely to be found in the smaller diameter ranges. Steel on the other hand, which makes up 4.4% of the total, is most likely to be used for large diameter and critical applications.

The American Concrete Pressure Pipe Association (ACPPA) carried out another investigation in 1997 in which a one-page survey instrument was distributed to around 110 utilities in the USA and Canada. This survey yielded the following information on installed pipeline materials; the respondents reported an average installed base of 1,141 miles of pipeline of all materials with responses ranging from as low as 10 miles to a high of 4,500 miles. The total installed pipeline base considered in the survey is therefore 63,495 miles, in a range of materials operating in utilities in Canada and America (Table 2.12).

The survey published data on the breakdown of materials for the size range 16-inch and above, which represents 4750 miles of pipeline.

Table 2.12
Pipe material breakdown ACPPA survey 1997, 16 inch and above

Pipe material	% of total
Cast Iron	32.00
Ductile Iron	28.20
PVC	2.20
Steel	3.40
Concrete Pressure Pipe	34.10

Source: ACPPA Pipeline Operations and Maintenance Cost Study 1997.

European Pipeline Population

Figures obtained by NWTC suggest that the potable water network in France amounts to some 838,000 km, (2002 statistics), and that the German network of water pipelines amounts to 281,000 km. This figure has been derived from a variety of sources; the supply situation in Germany is confusing since there are about 7000 private water distribution companies, which makes accurate network size predictions difficult and so the figures for the German networks have not been used in the subsequent analysis.

Predicted Joint Population

Having obtained and documented estimates of some of the water pipelines in service in the UK, France, USA and Canada, to get a feel for the numbers of joints present, it has been possible to arrive at an estimate. This has been achieved by using the average as-manufactured pipe lengths for the materials involved and dividing the estimated individual pipe material lengths by the appropriate average pipe length. The resultant figures for joint populations are presented in Table 2.13 and Table 2.14.

From the example, in excess of 300 million joints are likely to be present in pipelines carrying potable water supplies in some of the American, Canadian, UK and French water networks. This represents a considerable population of joints and on average, 70% of these are in old, deteriorating cast iron pipes. Even a small failure rate from these joints alone would be significant. A considerable potential market therefore exists for both internal and external joint repair, and joint location technologies. Knowledge of joint deterioration could be even more pertinent in areas with seismic activity.

Table 2.13
Estimated joint population by pipe material (from figure 2.11 - around a third of the US pipe network)

Pipe material	Joint population (millions)
Grey cast iron	49.9
Ductile cast iron	23.8
PVC	11.3
PE	0.35
Steel	2.2
GRP	0.11
Pre-stressed concrete	5.6
Asbestos cement	22.7
Others	2.1
Total	118.06

Table 2.14
Estimated joint population by pipe material (from above USA/figures plus UK/France figures)

Pipe material	Joint population (millions)
Grey cast iron	128
Ductile cast iron	61.1
PVC	66.1
PE	2.25
Steel	3.6
GRP	0.1
Pre-stressed concrete	5.6
Asbestos cement	35.2
Others	24.9
Total	327.75

JOINT FAILURE DATA

Incidences of joint failure compared to other causes have been obtained from a number of sources. Details are presented below for information. Data source, National Research Council (NRC), Canada. Table 2.15 has been reproduced from the results of a study of water main failure modes in 21 Canadian cities during 1992 and 1993 related to pipe materials.

Table 2.15
Mains breakage history - NRC study

Pipe material	Year of investigation	Number of breaks	Pipe defects		Joint failures		Other causes	
			No	%	No	%	No	%
Cast Iron	1992	3075	2822	92	179	6	74	2
	1993	3216	2897	90	197	6	122	4
Ductile Iron	1992	392	374	95	15	4	3	1
	1993	414	390	94	7	2	17	4
Asbestos	1992	114	93	82	3	3	18	15
Cement	1993	129	105	81	11	9	13	10
PVC	1992	17	9	53	5	29	3	18
	1993	9	2	22	3	33	4	45
Concrete	1992	3	Na		Na		Na	
	1993	5	Na		Na		Na	

Source: National Research Council (NRC), Canada 1993.
Note: Na = not available

The above data shows the scale of cast iron joint problems with 6% of all failures attributed to joints (around 180 joints per year). Work carried out by the NRC in Canada has also identified a particular problem with pipe joints in spun iron laid in the late 1940's, in which a sealing material called Leadite was used. This material was sulphur based and has a different thermal expansion from the spun iron. Very low and high temperatures can therefore give rise to changes in the volume of the sealing material leading to cracks in the bell of the pipe. This can lead to sudden failure of the bell and immediate leakage from the joint.

Additional data has been obtained from Essex and Suffolk Water (UK) where over a one-year period, the failure modes of three pipe materials (cast iron, spun iron and asbestos cement) were investigated. This study produced the results shown in Table 2.16. In both of the studies discussed here, the nature of the joint defect was not elaborated upon, but simply classed as a failure. Hence, it could be assumed that leakage had occurred at a progressive rate until it could be detected and at which point it would have been recorded in the study.

Table 2.16
Essex and Suffolk data

Pipe material	Pipe defects % total	Joint defects % total
Cast Iron	97.00	3.00
Spun Iron	90.00	10.00
Asbestos Cement	90.00	10.00

Source: Northumbrian Water Ltd.

CHAPTER 3
UTILITY SURVEY

INTRODUCTION

This section presents the findings from the Utility Survey. Response to questionnaires was not as good as expected, with only seven returns. Therefore, in view of the limited size of the data set considered, caution should be exercised in the interpretation of results. The results only provide useful indications of the numbers, types and perceived causes of joint failures from a range of water utilities in the USA, Canada and Europe.

DEVELOPMENT OF QUESTIONNAIRE

The questionnaire was designed to assist in developing a better understanding of joint failure problems faced by the water industry. The type of information requested included:

- Mains population by pipe material and diameter
- Joint types used by pipe material
- Modes of joint failure (type and cause)
- Joint condition assessment undertaken
- Methods of joint rehabilitation used

The questionnaire was circulated to AwwaRF, the utility partners and PAC members for review, updated in accordance with comments received and approved for circulation by AwwaRF in May 2002.

Information from the survey assisted in targeting the most appropriate pipe materials and joint types (most numerous population) on which to focus investigations of failure mechanisms, and joint location, condition assessment and repair techniques.

SURVEY RESPONSE

Completed questionnaires were received from seven companies:

- Essex and Suffolk Water (UK)
- United Water (New Jersey, USA)
- Charleston CPW (South Carolina, USA)
- City of Calgary Waterworks (Calgary, Canada)
- Aqua Pennsylvania Inc. (Philadelphia, USA)
- City of Ottawa, Water Distribution (Ontario, Canada)
- East Bay Municipal Utility District (Oakland, USA)

In order to maintain an element of confidentiality, the data has been presented by reference to utility A, B etc, assigned randomly to the above utilities. This assignment does not detract from the survey or the findings, as the purpose is to generally understand the main problem areas for utilities associated with pipe joints, and not individual utility requirements. The questionnaire itself was not fully completed by all companies. Table 3.1 shows the sections of the questionnaire completed by each responding company.

Table 3.1
Sections of questionnaire completed

		Company name						
Section no.	Section description	Utility A	Utility B	Utility C	Utility D	Utility E	Utility F	Utility G
1	General Information	•	•	•	•	•	•	•
2.1	Network Length & Age	•	•	•	•	•	•	•
2.2	Operating Conditions	•	•	•	•	•	•	•
3	Jointing Techniques	•	•	•	•	•	•	•
4.1	No. of Joint Failures	•	•	•	•	•	•	•
4.2	Joint Failure Mechanisms		•	•	•	•	•	
4.3	Reasons for Joint Failures		•	•	•	•	•	•
5.1	Location of Joints		•	•	•	•	•	•
5.2	Joint Condition Assessment		•	•	•	•	•	•
5.3	Joint Repair Methods		•	•	•	•	•	•
6	Cost of Joint Repair		•	•	•	•	•	
7	Supply Interruptions		•	•	•	•	•	

Data and findings from the utility survey are presented in the following Sections. Table 3.2 shows abbreviations used for different materials.

Table 3.2
Abbreviations used for material types

Pipe material	Abbreviation
Cast iron	CI
Ductile iron	DI
Glass reinforced plastic	GRP
Polyethylene	PE
Poly vinyl chloride	PVC
Pre-stressed concrete	PSC
Reinforced concrete	RC
Stove-cement	SC
Steel	Steel

SURVEY RESULTS

Mains Population

The total water mains population considered in this survey (i.e. sum of data received from all companies) was 20,779 miles (33,247 km). This population is broken down by pipe material and diameter in Figure 3.1, and tabulated in Table 3.3.

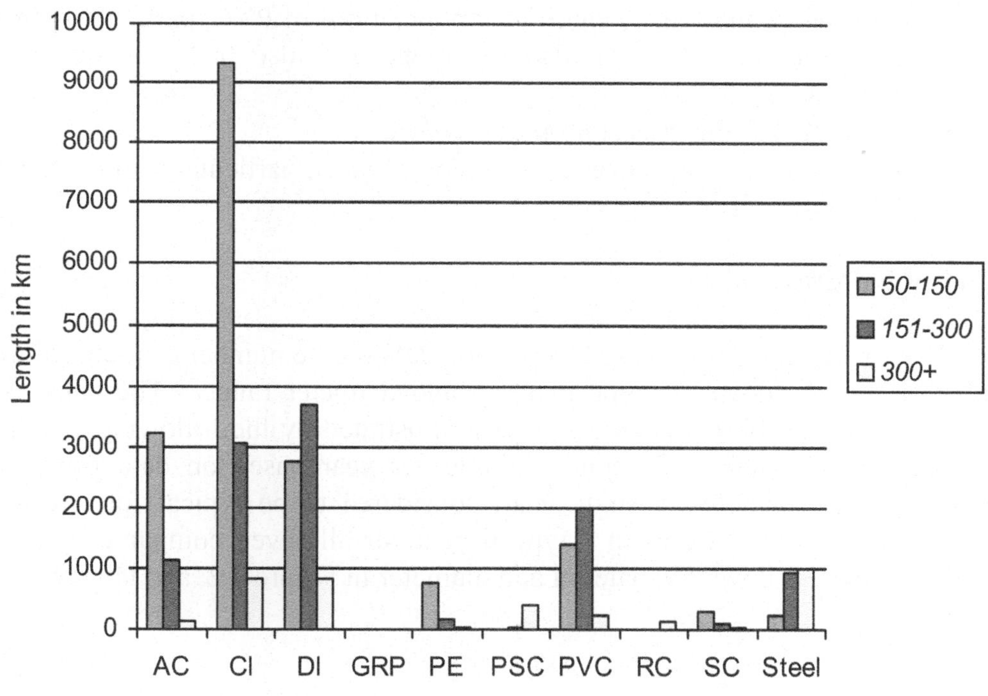

Figure 3.1 Mains population by material and diameter

Table 3.3
Mains population (length in metres) by material and diameter

Material	Diameter range (mm)		
	50 – 150	151 – 300	300+
AC	3,210	1,127	117
CI	9,347	3,048	821
DI	2,749	3,700	846
GRP	2	7	16
PE	768	168	21
PSC	2	45	390
PVC	1,389	2,004	221
RC	1	1	117
SC	306	94	24
Steel	243	945	991
Totals	18,017	11,139	3,564

The mains population for each company is broken down by material and diameter in Appendix A. The following observations can be made:

- CI and/or DI are generally dominant
- Utilities B and G have large populations of AC in diameters up to 12 inch (300 mm)
- Utilities A and F have significant populations of PVC, particularly in the 6 to 12 inch (150 mm to 300 mm) diameter range and also have the highest proportions of PSC/RC
- Utility B has the largest proportion of PE
- Utility G has a significant population of steel, particularly in diameters above 6 inch (150 mm)

Number of Failures

Water companies were asked to provide data on the number of joint failures recorded in a typical year, broken down by pipe material and diameter range. The data provided by each company was from different sources. This is illustrated by the following examples: example 1, water mains break database, average failures per year based on data from the last 5 years; example 2, failure data for a single year, considered to be typical for any given year. The combined sum of joint failures in a typical year for all seven companies was 603. This joint failure data is broken down by material and diameter in Figure 3.2, and tabulated in Table 3.4.

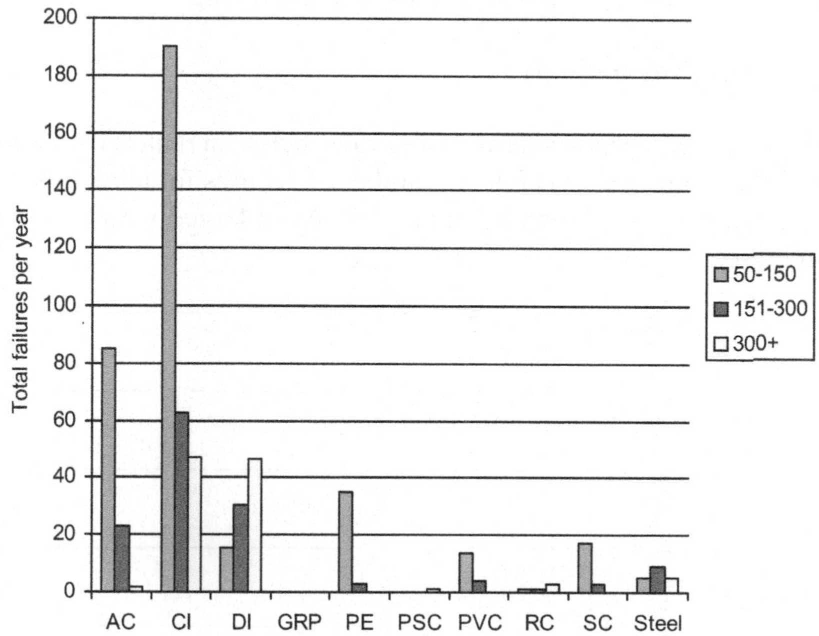

Figure 3.2 Total numbers of failures in a typical year by material and diameter

Table 3.4
Total numbers of failures in a typical year by material and diameter

Material	Diameter range (mm)		
	50 – 150	151 – 300	300+
AC	85	23	2
CI	190	63	47
DI	15	31	46
GRP	0	0	0
PE	35	3	0
PSC	0	0	1
PVC	14	4	0
RC	1	1	3
SC	17	3	0
Steel	5	9	5
Totals	362	137	104

The total number of failures in a typical year is broken down by company, material and diameter range in Appendix B. The following observations are made:

- CI shows the highest number of failures per year in all diameter ranges and for all companies.
- Utility B has the highest number of failures for AC, CI and PE in the 2 to 6 inch (50 to 150 mm) diameter range.
- Utility E has the highest number of failures for CI and DI in the 6 to 12 inch (151 to 300 mm) and the 12+ inch (300+ mm) diameter range.

- There were no reported failures for the 25km length of GRP pipe

Number of Failures (Normalised)

In order to allow comparison of failure rates between materials, failures have been normalised to failures/km/year. The total number of failures for all companies per km per year, by material and diameter are shown in Figure 3.3. An enlarged view is provided in Figure 3.4 so that detail can be seen.

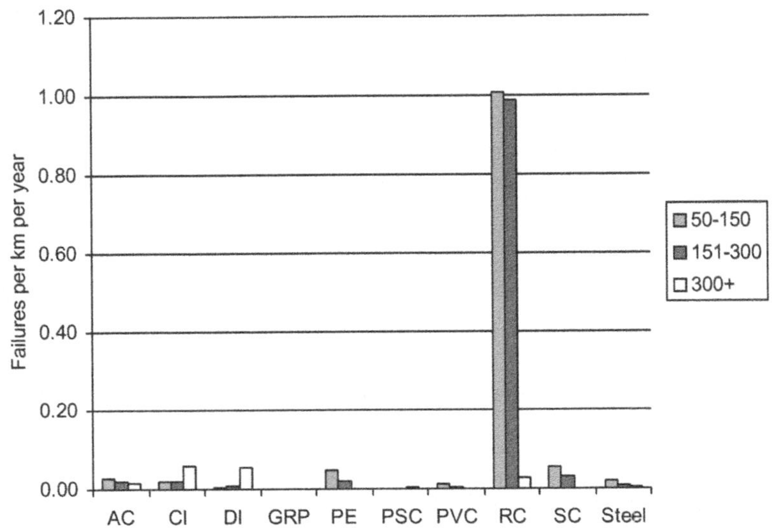

Figure 3.3 Failures per km per year by material and diameter type

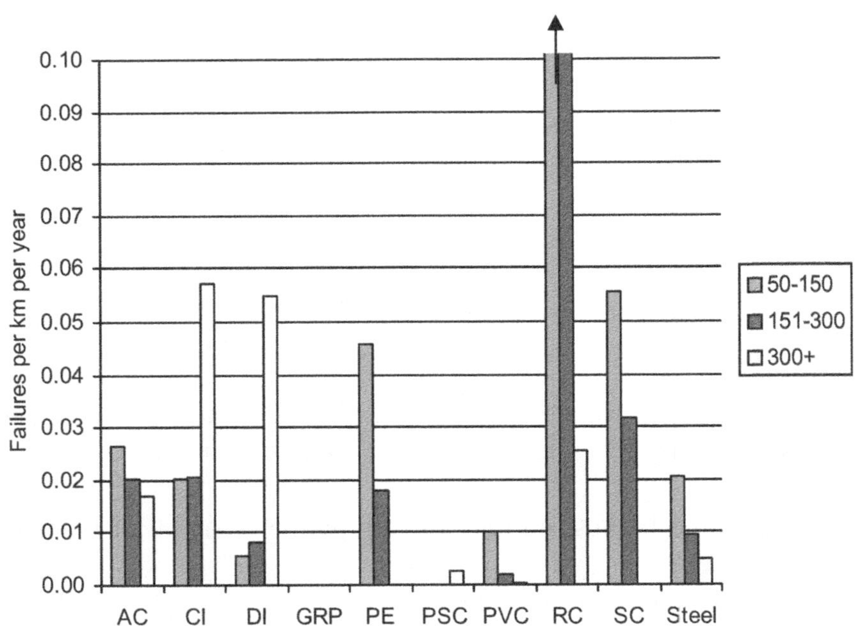

Figure 3.4 Enlarged view of Figure 3.3

The following observations can be made:

- RC dominates because of a few failures in a small network length. This is not statistically valid.
- CI and AC exhibit high failure rates over the full range of pipe diameters.
- The high failure rate for small diameter PE is biased by Utility B data as explained below.

Charts of the number of failures normalised to failures/km/year are shown in Appendix C for each of the diameter ranges.

Types and Causes of Joint Failure

The types and causes of joint failure for each pipe material are shown in Figures 3.5 to 3.20 below. For example, Figure 3.5 shows the percentage of total cast iron joint failures (for all responding companies) by failure type, and Figure 3.6 shows the percentage of total cast iron joint failures by failure cause. This method of representing the data is not statistically robust, since failure types and causes for the entire data set for each material are represented on single charts, and there are instances where circumstances particular to a single company's data may influence results. However, representing the data in this way does allow some general observations to be made. Some mis-assignment of information is also apparent for example in Figure 3.12, gasket degradation has been given as a cause of failures on PE pipes. There would not normally be any gaskets present on PE pipes unless a flange adaptor was used to connect to another fitting such as a valve.

Failure Types – Main Observations

- The dominant type of joint failure (50% - 100%) for all mechanically jointed pipes was gasket/seal failure. This includes CI, DI, AC, steel, PVC and RC.
- For CI, DI and AC, bolt failures account for around 20% of total failures.
- CI exhibits a significant percentage (18%) of structural socket failures.
- PE joint failures are dominated by electrofusion coupling failures, however, this is based on data from Utility B only, and was primarily caused by incorrect installation.
- PVC joint failures were dominated by gasket/seal failure (55%) but with a large proportion of structural socket failures (40%).
- All stove-cement joint failures were structural spigot failure.

Failure Causes – Main Observations

- The primary cause of joint failure for mechanically jointed metallic pipes (CI, DI) was gasket degradation.
- The primary causes of joint failure for mechanically jointed non-metallic pipes (AC, RC, PVC) was ground movement

- Mechanically jointed steel pipe exhibited equal proportions of gasket degradation and ground movement.
- A large proportion of failures for RC (35%) are caused by corrosive ground.
- The primary cause of stove-cement joint failures (50%) was pressure surge.

Appendix D gives percentages of joint failures by type and cause for each of the five companies completing these sections of the survey. This allows company specific trends to be considered.

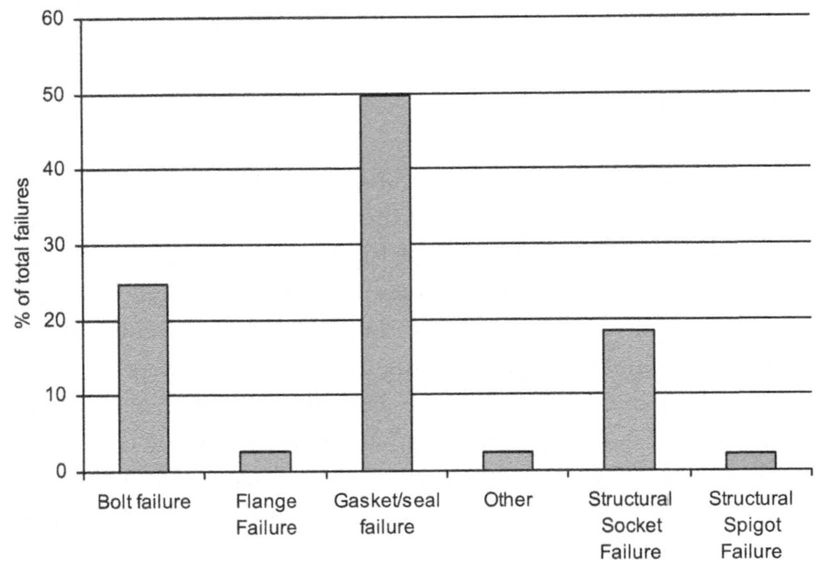

Figure 3.5 Failure types – CI

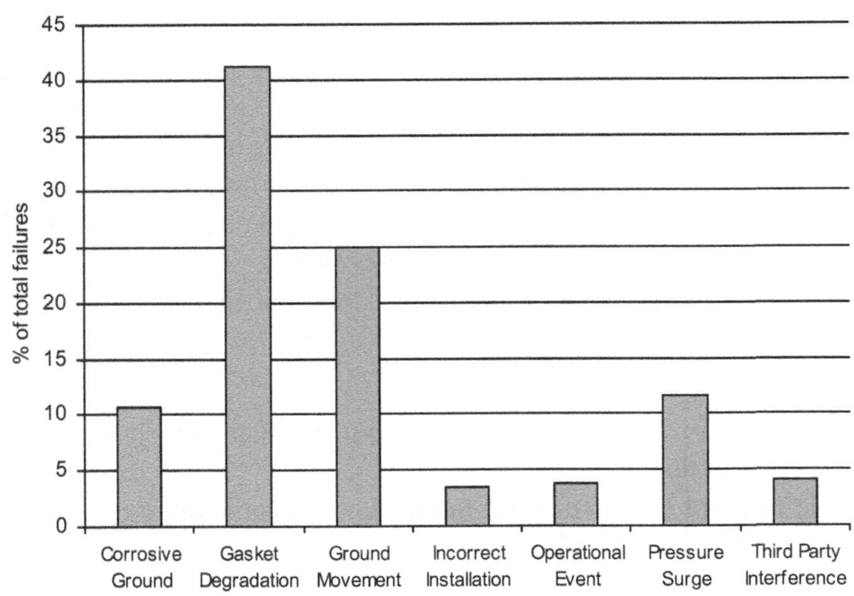

Figure 3.6 Failure causes – CI

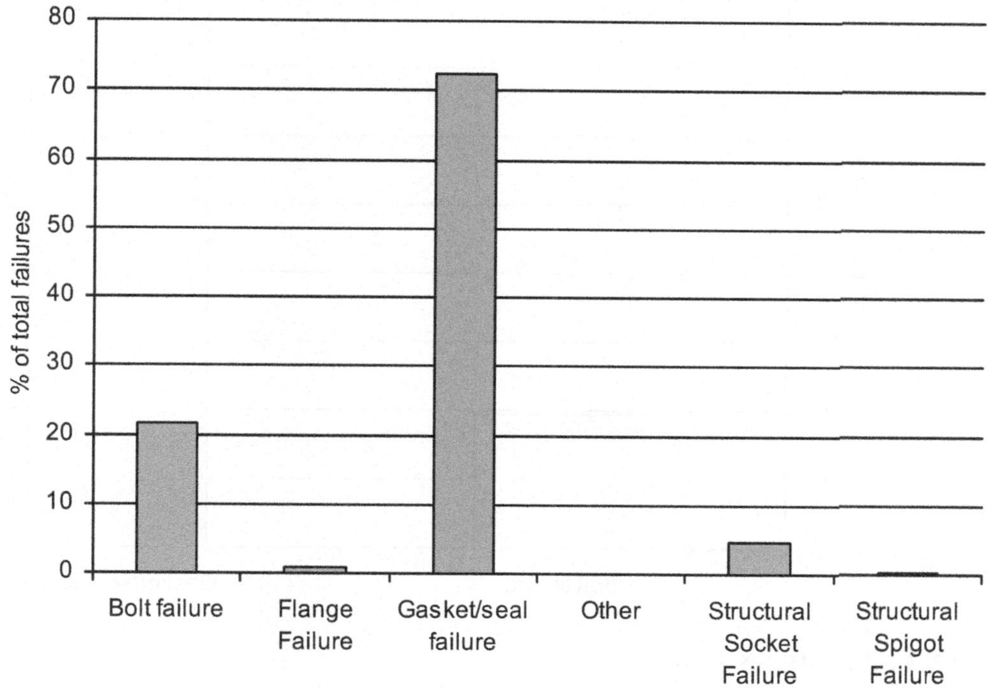

Figure 3.7 Failure types – DI

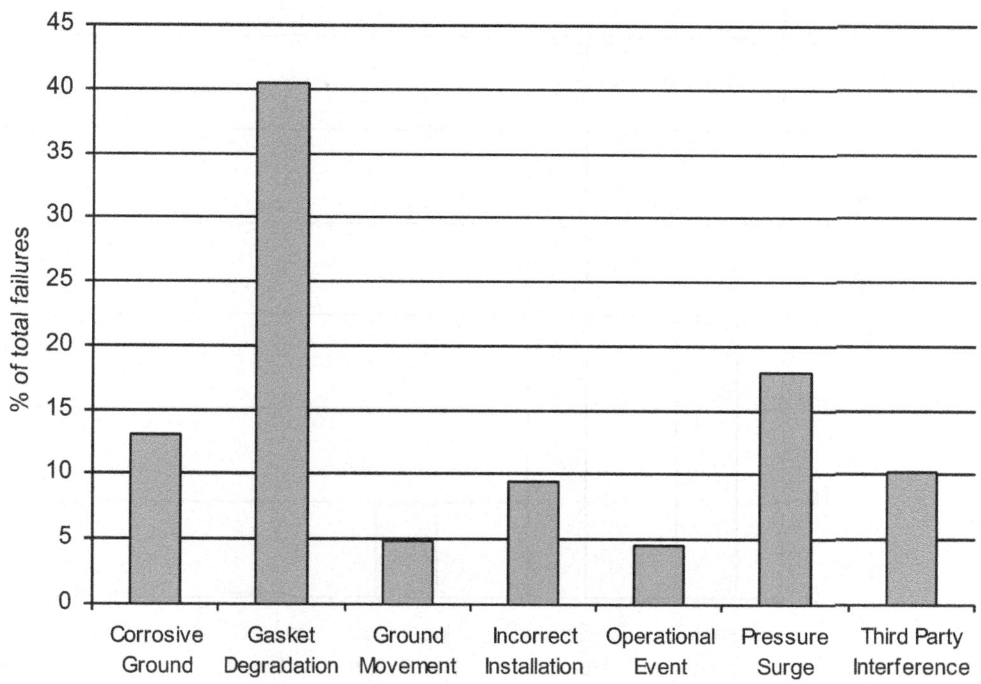

Figure 3.8 Failure causes – DI

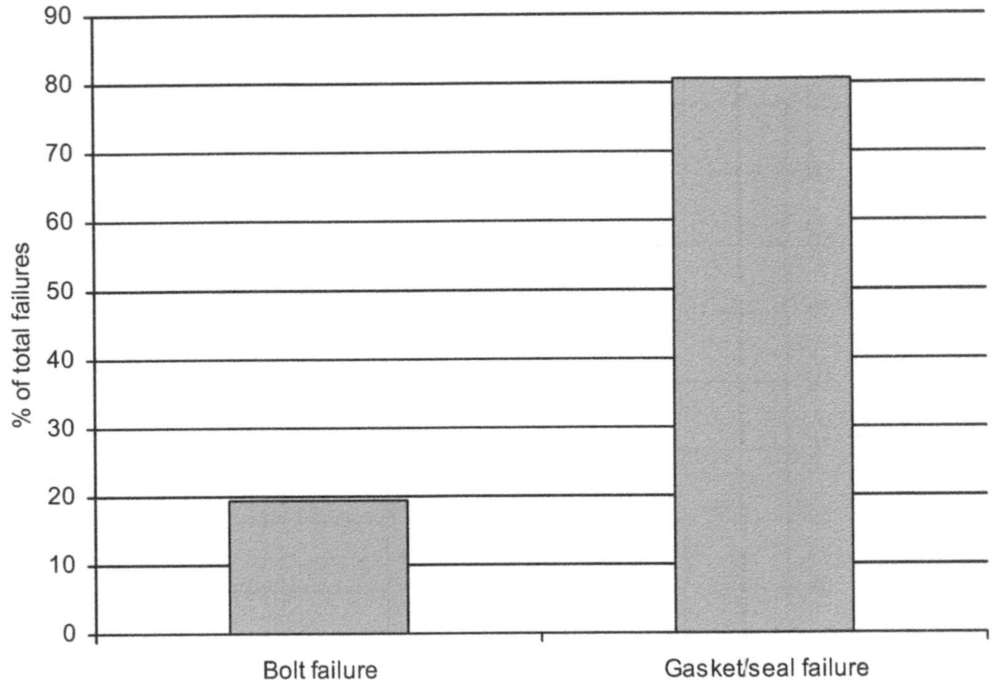

Figure 3.9 Failure types – AC

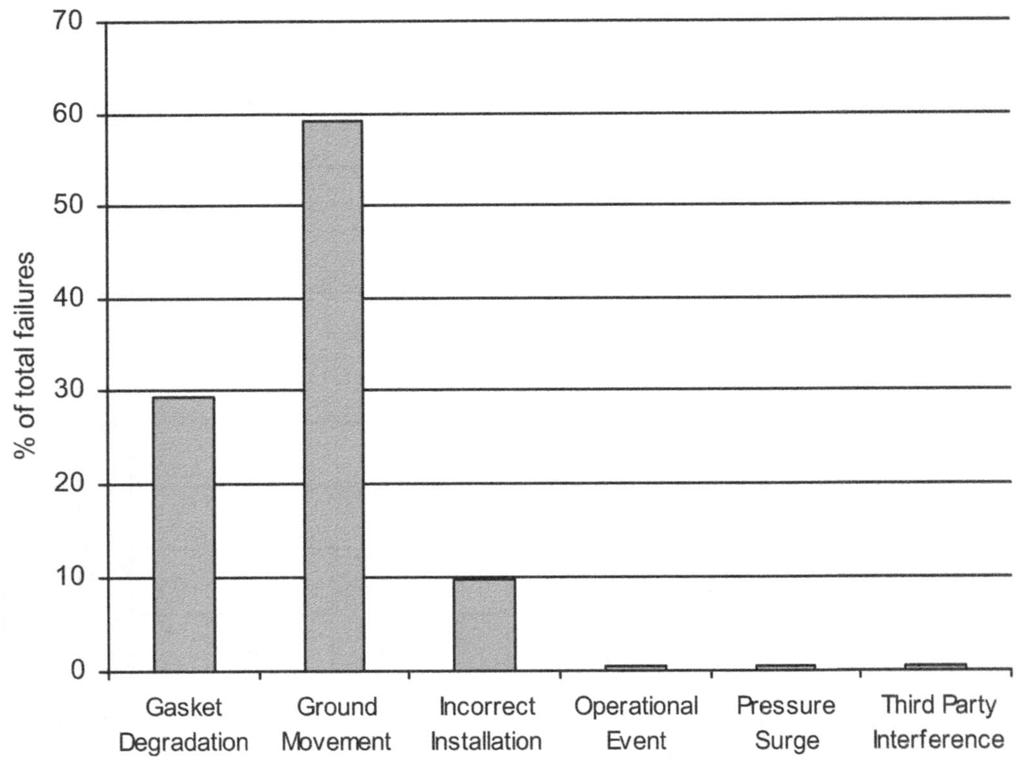

Figure 3.10 Failure causes – AC

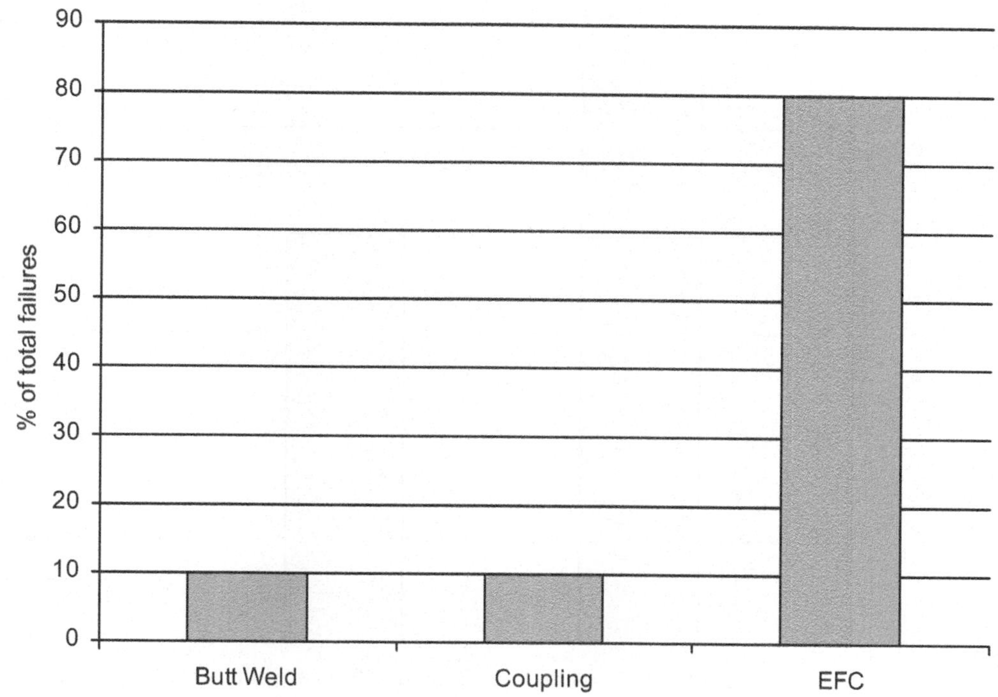

Figure 3.11 Failure types – PE

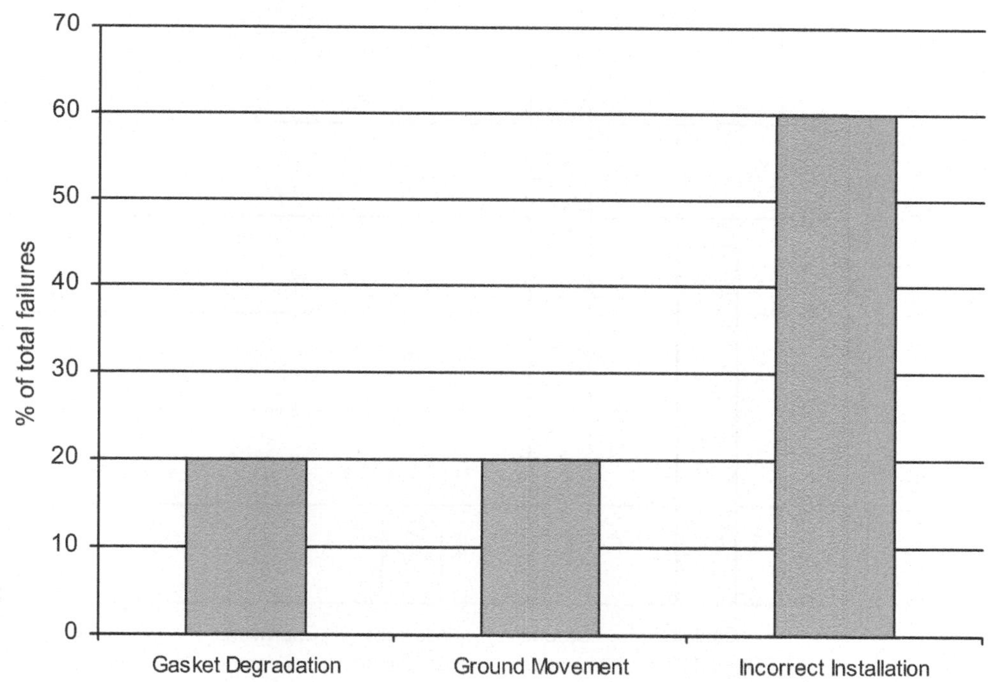

Figure 3.12 Failure causes – PE

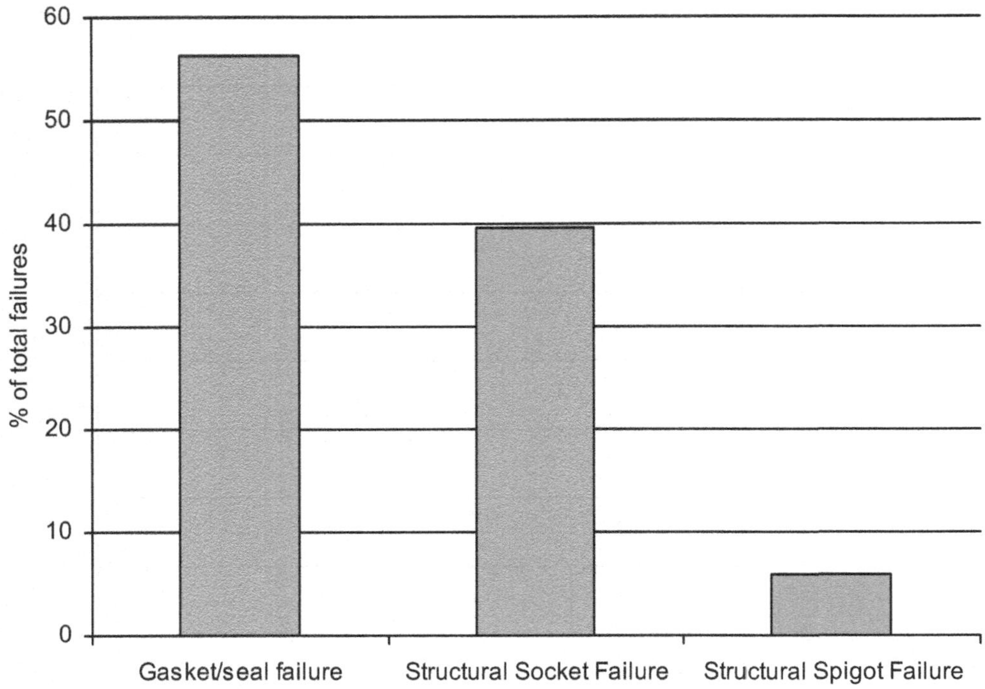

Figure 3.13 Failure types – PVC

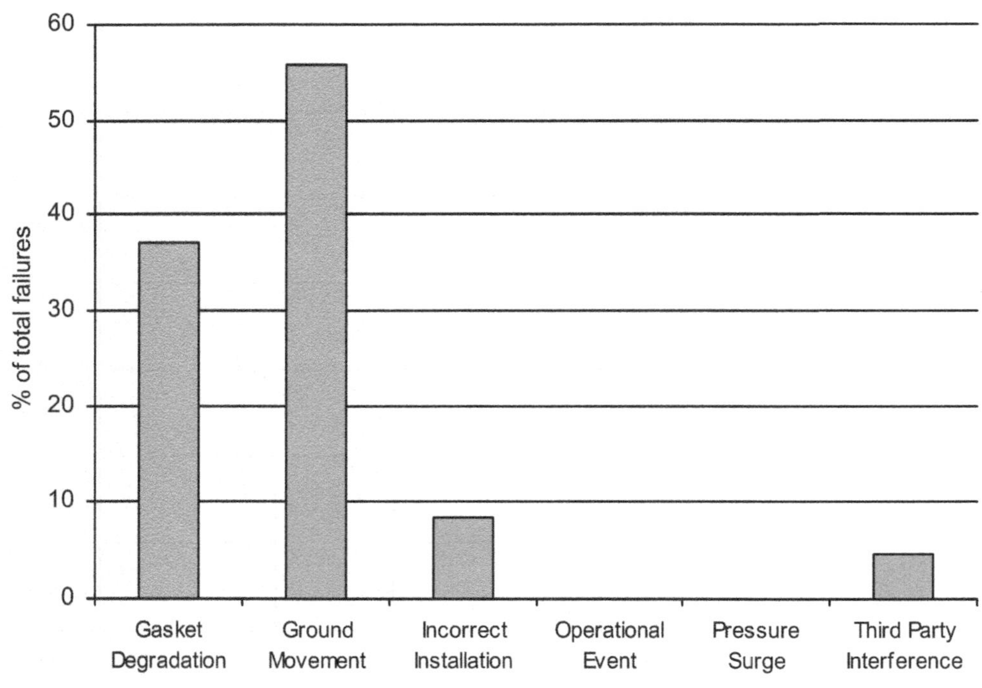

Figure 3.14 Failure causes – PVC

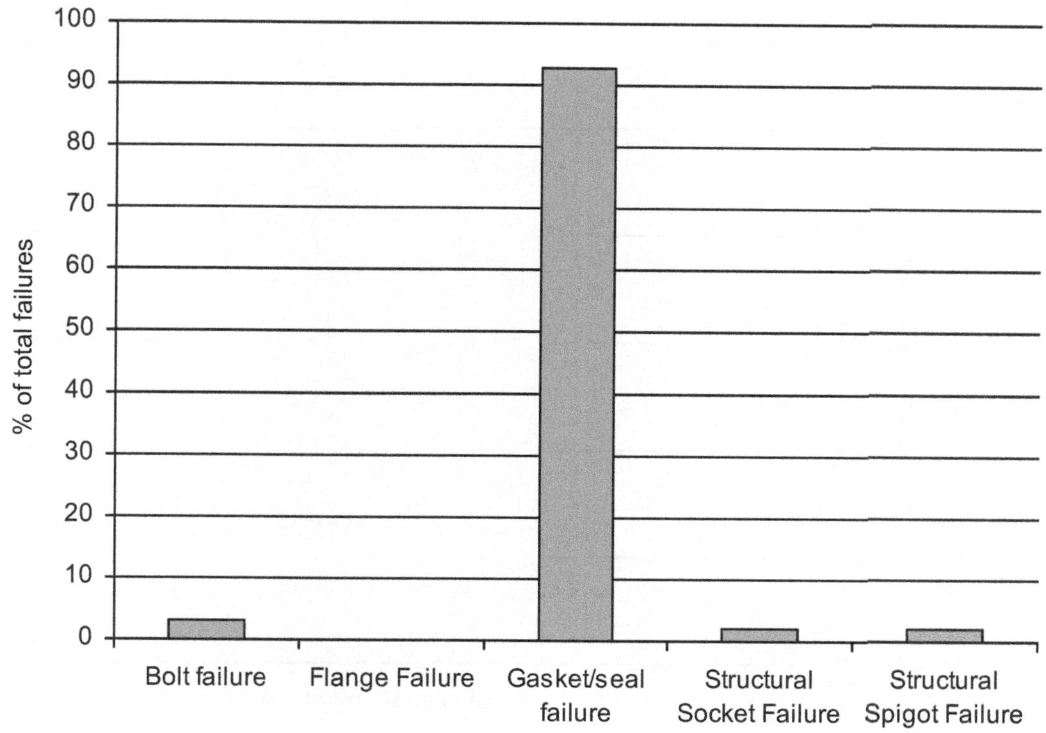

Figure 3.15 Failure types – Steel

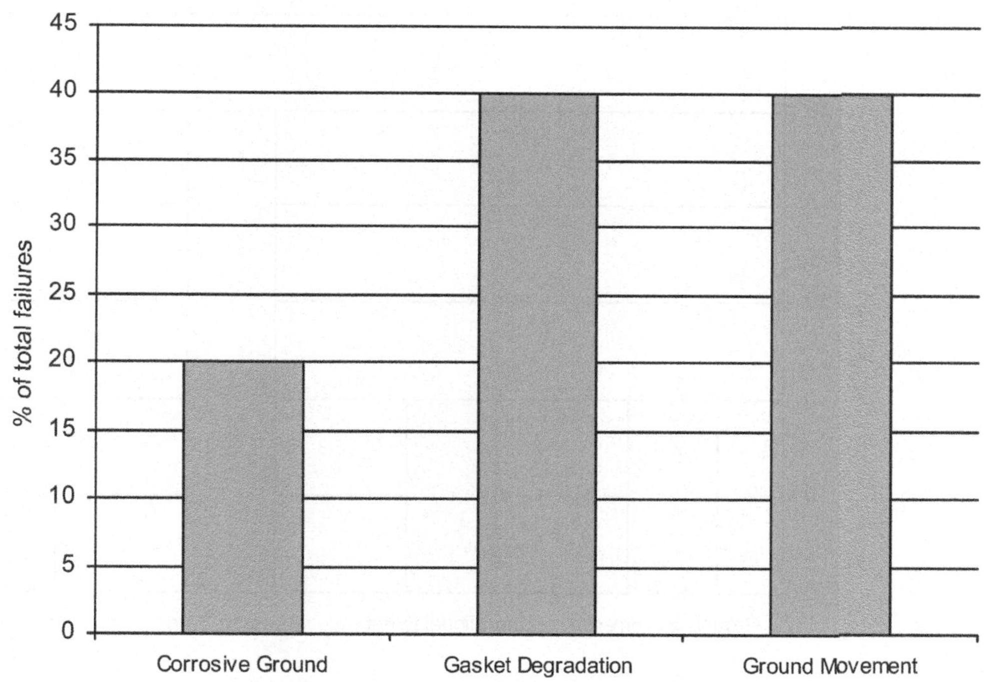

Figure 3.16 Failure causes – Steel

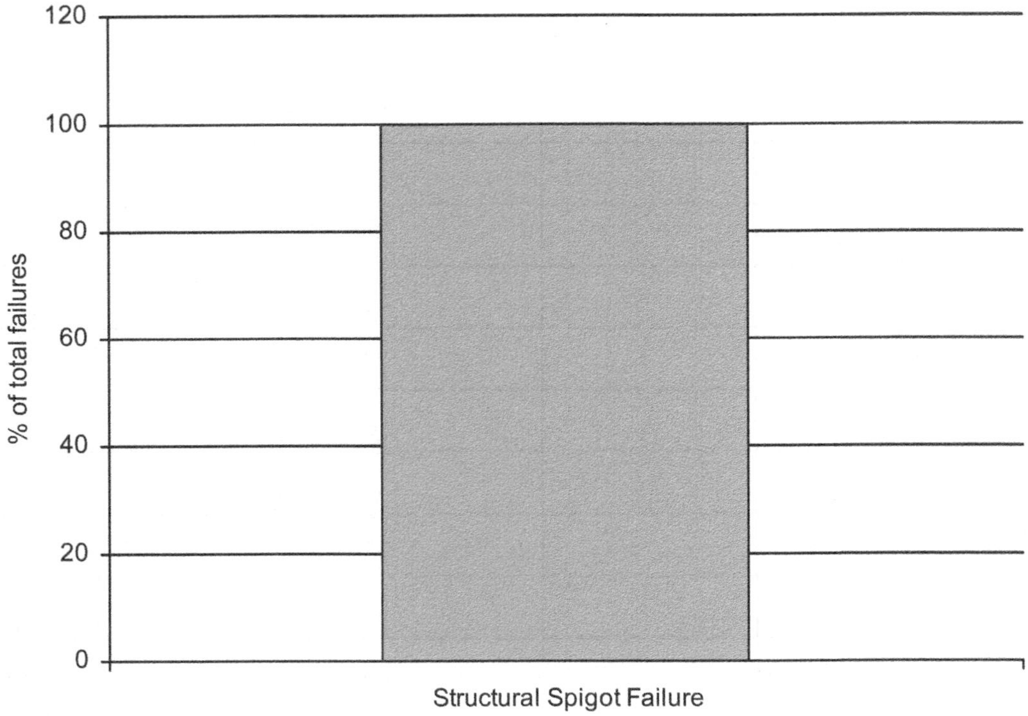

Figure 3.17 Failure types – SC

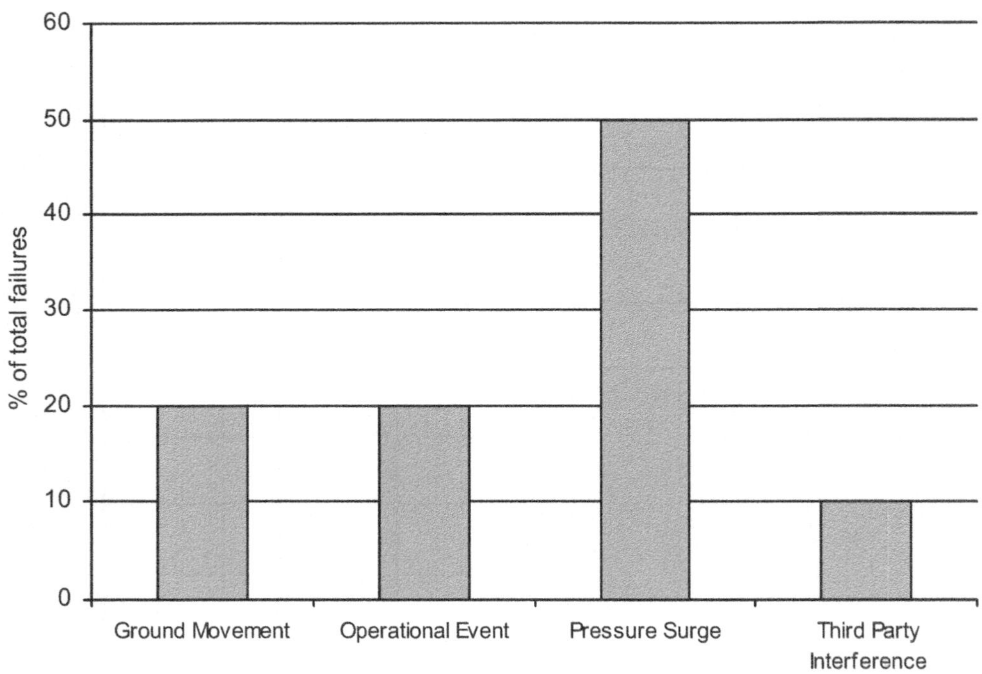

Figure 3.18 Failure causes – SC

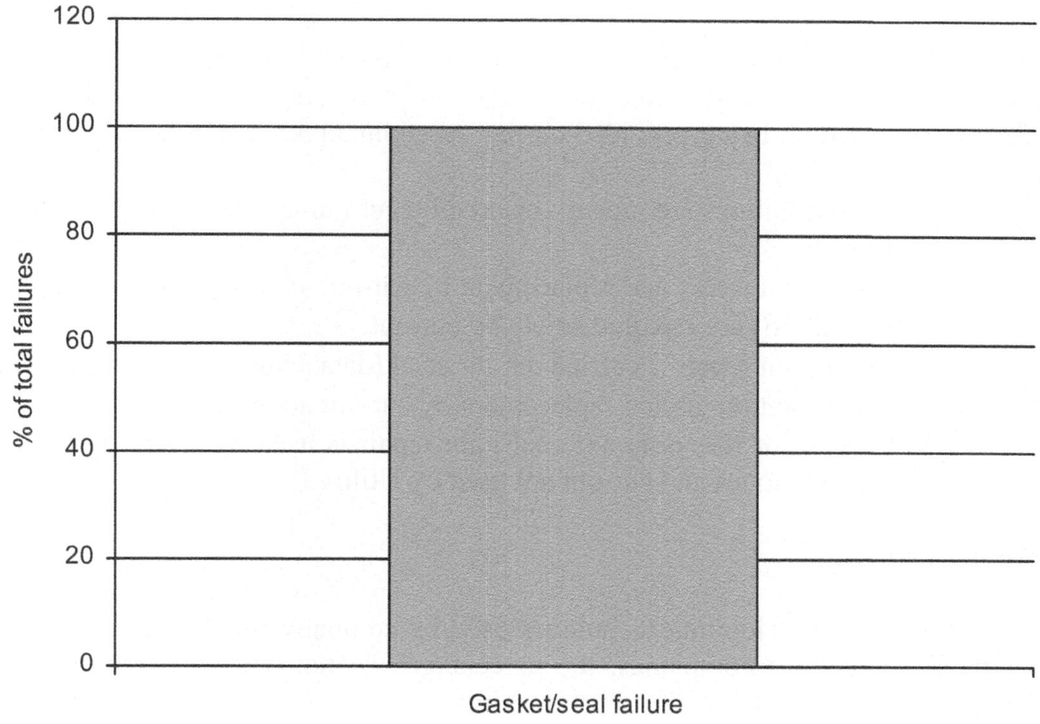

Figure 3.19 Failure types – RC

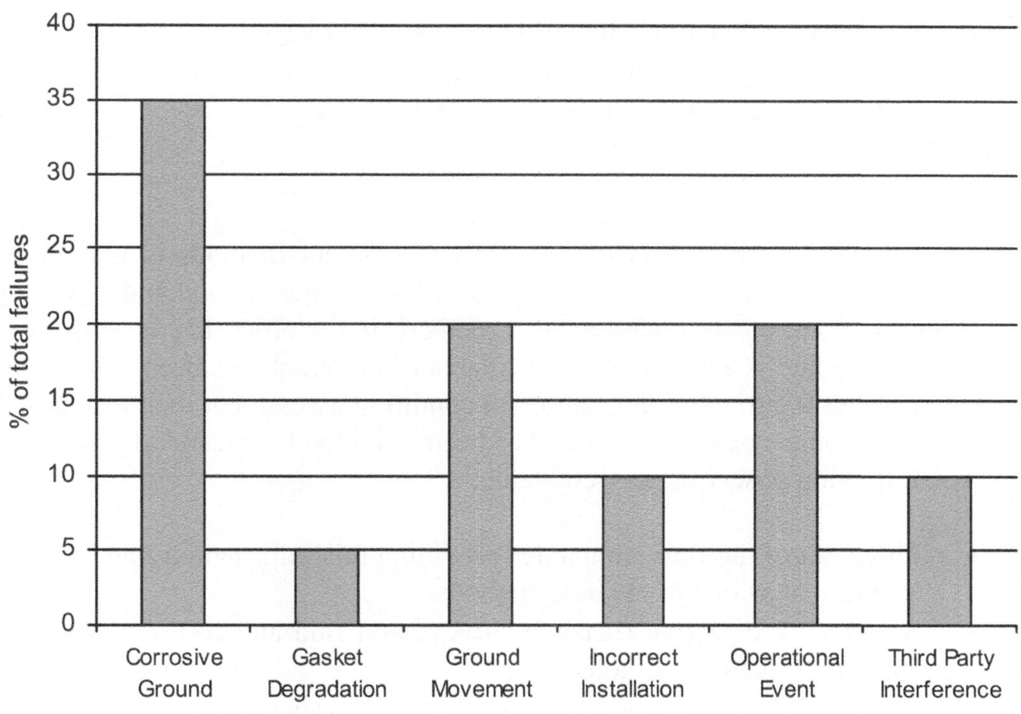

Figure 3.20 Failure causes – RC

Joint Repair Methods

Joint repair methods used by each company and for each of the main pipe materials are shown in Appendix E. For some companies, the percentages do not add up to 100 due to insufficient information being provided on the questionnaires. The following observations were made:

- CI and DI failures are mostly repaired by cut-out and replacement of the affected pipe section.
- External joint repair and replacement by cut-out are used in equal proportions for AC.
- PVC (and PE) are repaired solely by cut-out.
- External joint repair is carried out for steel (data available only from Utility B).
- External joint repair and replacement by cut-out are used for PSC.
- In the case of RC, only external joint repair is used by Utility F while both internal lining techniques and cut-out are used by Utility E.

Jointing Techniques

A breakdown of jointing techniques used by company for each pipe material is given in Appendix F. For some companies, the percentages do not add up to 100 due to insufficient information being provided on the questionnaires.

- CI joints are mainly Push Fit (lead/yarn) and Bolted Flange
- DI joints are mainly Push Fit Spigot & Socket
- Steel joints are mainly Butt-weld and Welded Sleeve
- PE joints are mainly Butt-weld except for Utility D where electrofusion is used.
- PVC joints are mainly Push Fit Spigot & Socket.

Joint Condition Assessment

The reported effectiveness of condition assessment methods is graphically depicted in Figure 3.21. A scale from 1 to 3 is used representing *very good*, *good* and *poor* respectively. The legend shows the joint condition assessment methods in the order they appear for each company. If no particular method is employed then that area of the graph has been left blank. Although the utilities were asked to provide information on condition assessment methods used for joints, it is suspected that in some cases, information has been provided from general pipe condition surveys.

The following observations were made:

- Three out of the four companies providing this information rated excavation & visual assessment as one of the best methods.
- Active leak detection (leakage surveys and find and fix) and pressure/flow surveys are other effective methods.
- Internal acoustic leak detection has varying levels of success from *very good* to *poor*.
- Ultrasonic methods have a *very good* rating but are used only by Utility E.

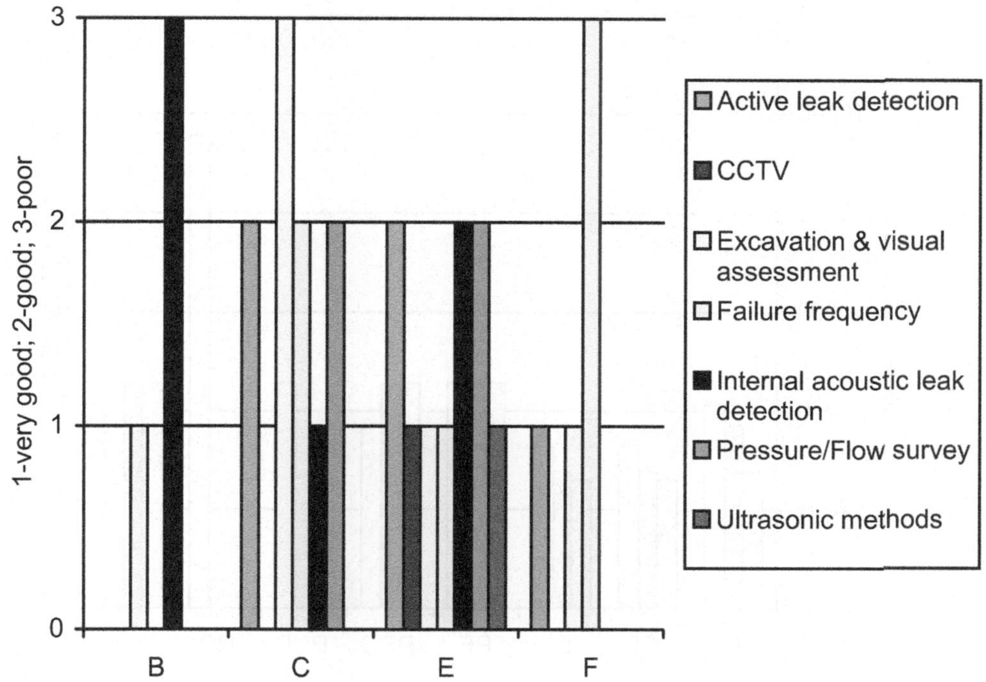

Figure 3.21 Effectiveness of joint condition assessment methods

Cost of Joint Repair

Figure 3.22 shows the mean cost of joint repair (US Dollars) for all companies responding, broken down by material and diameter range. These costs include labour and direct costs.

The costs were not broken down consistently by all companies responding and have therefore been combined (labour and direct costs) to give a total cost of joint repair on a best merits basis. It is considered that these costs should be used as a general guide only. The combined cost of joint repair as provided by each responding company is shown in Appendix G. The exchange rate of 1 CAD to 0.76 USD (as of 31st October 03) was used. Utility G provided an average joint repair cost of $4000.

The following observations were made:

- Utility A has the highest joint repair costs in all diameter ranges. This is due to the high cost of excavation (deep pipes).
- After Utility A, Utilities E and C have a higher repair cost than the other companies in all diameter ranges.
- Utility D has the lowest repair costs for CI, DI and PVC in all diameter ranges.
- The differential between diameter ranges is not as significant as may have been expected.

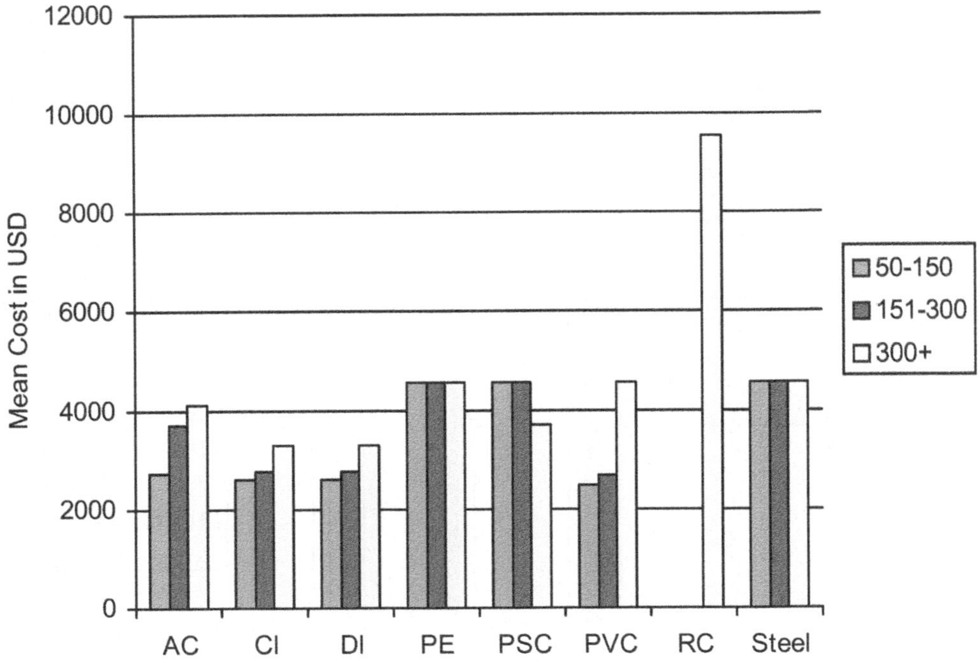

Figure 3.22 Mean cost of joint repair by material and diameter

Sources of Information

Information sources were requested on the questionnaires, in order to provide an indication of likely accuracy and reliability. The sources of information for the available data, as received on the questionnaires, are tabulated in Appendix H. These have been grouped into 4 classes as depicted in Table 3.5

Table 3.5
Sources of information

Source description	Class
Databases/spreadsheets/operating reports/records	1
Personal/local/system knowledge	2
GIS	3
Interviews with management	4

Figure 3.23 shows the distribution (expressed as a percentage) of the sources of information. The large majority of information sources are databases and personal knowledge.

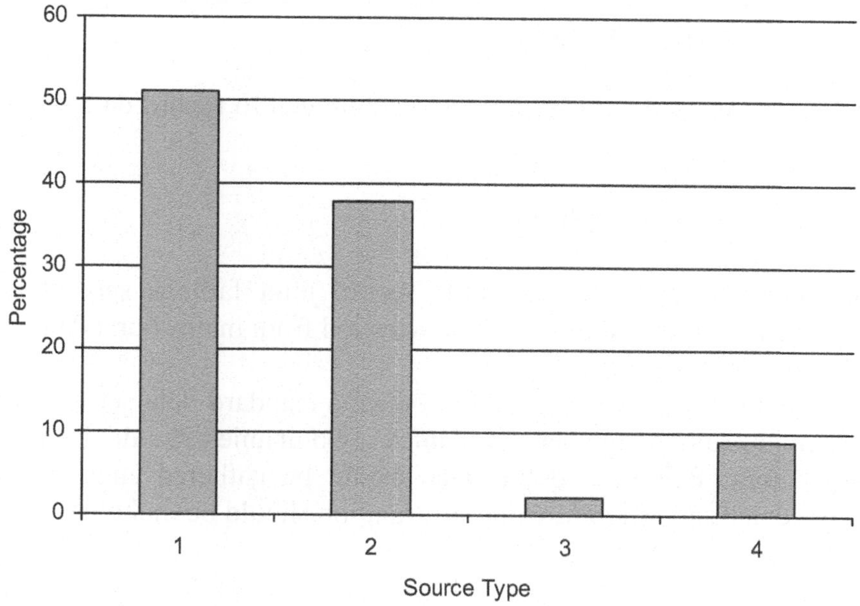

Figure 3.23 Distribution of information sources

It should be expected that information from databases etc. is reliable and of good quality, however, experience shows that this is not always the case. Information obtained by word of mouth, e.g. discussions with personnel, can be very accurate even though it is not formally recorded. Therefore, the questionnaire does not provide a clear indication of the quality and reliability of data.

CONCLUSIONS FROM UTILITY SURVEY

The response to the Utility Survey was poor and it can be argued that the data set is not large enough to be statistically robust. The data provided on numbers of joint failures was for a typical year. Some companies provided this for a single year, considered to be typical of all years, and others provided data averaged over a number of years. It is recognised that where small numbers of joint failures or short lengths of pipeline are considered, then inaccuracies in the data provided can have a significant affect on calculated results. To improve confidence in the estimates of joint failure rates, it would be necessary to acquire failure data from a much larger sample of companies and/or to acquire data over a long time history. Subject to the above comments, the data does provide useful information on:

- Numbers of joint failures and joint failure rates by material and company
- The primary types and causes of joint failures in water pipelines
- Joint repair methods used
- Jointing techniques used for different materials
- Joint condition assessment techniques used
- Mean costs of joint repairs by material, diameter and company

- Joint failure rates, which were perhaps high where expected but also highlighted specific local problems

Water companies should be actively encouraged to record data on joint failures. This is discussed below.

Data set for recording joint failures

Water companies do not generally record joint failures separately from other failure information. Data on joint failures is often extracted from mains burst data, work records etc. and this can be a laborious process.

It is recommended that AwwaRF produce a standard data set for recording joint failures and circulate this around member companies. Companies should be actively encouraged to record data on joint failures and the data should be gathered annually and compiled into a comprehensive database. The following information should be included:

- Date of failure
- Geographic location (map reference or street and house number)
- How the joint failure was detected (e.g. reported leak, active joint condition assessment)
- Pipeline material
- Pipeline diameter
- Joint type (generic types as covered in this report)
- Age of pipeline (date the pipeline was installed)
- Location of pipeline (major road, minor road, pavement, grass verge, etc.)
- Burial depth
- Ground type (details of backfill, water content etc.)
- Failure type (generic types as covered in this report)
- Failure cause (this may be personal judgement but can still be useful)
- Evidence of previous repairs (has the joint or parent pipe been repaired before)
- Remedial action taken (repair method employed)
- Cost of repair

Over time, this would provide a comprehensive database on joint failures. Analysis of the data would provide asset managers with accurate information on the scale and significance of joint failure problems, and engineers with a greater understanding of the types and causes of joint failures and the requirements for joint condition assessment technology.

CHAPTER 4
UNDERSTANDING JOINT BEHAVIOUR

JOINT PARAMETERS

Joint Sealing Performance

Due to the large number of potential parameters associated with joint degradation mechanisms, a review of the understanding of these important factors was undertaken. Most of the different pipe materials and joint types were considered in turn, with regard to what were considered to be possible modes of degradation and subsequent failure. The following parameters were considered to be the most representative in influencing joint condition and possible failure for all pipe materials.

Parameters contributing to degradation of all joint types:

- Angular deflection
- Axial pull
- Backing ring corrosion
- Bolt corrosion
- Cracking of spigot or socket
- Differential expansion of joint materials
- Embrittlement
- Gasket failure
- Gasket relaxation
- Gasket/seal degradation
- Incorrect dimensions e.g. insert size
- Internal and external corrosion
- Lateral displacement
- Mixed/incompatible materials
- Ovality
- Pipe material degradation
- Poor joint construction (e.g. adhesion)
- Stress at joint junction
- Weld quality
- Degradation consequence – leakage

For consistency, it was decided that joint condition would be ranked as a level or class of degradation. This is explained more fully in the section on joint degradation. For each of the materials and joint types identified earlier in this chapter, all known degradation/failure modes were detailed. An example of the entries for cast iron pipes is shown in Table 4.1 and the full summary for all materials is shown in Appendix I. Consideration was also given to possible inspection techniques, existing, emerging, innovative etc. for each degradation mode by either internal or external means where applicable.

Table 4.1
Example of entries for cast iron pipe

Material/joint type	Degradation mode	Internal technique	External technique
Cast iron Push-fit Lead/yarn	Angular deflection	Measure width of joint gap around circumference using – caliper pig or magnetic sensors Inertial Navigation System pig (INS)	Manual measurement Global Positioning System Ground Probing Radar
	Axial pull out	Measure width of joint gap around circumference using – caliper pig or magnetic sensors	Visual examination of spigot
	Lead/yarn deterioration	Core sample of lead and yarn	Core sample
	Lead slump	None	None
	Thermal expansion/ Cracked spigot	Ultrasonic (US) pig	None
	Lateral pipe movement	Caliper pig - Measure eccentricity between spigot and socket	Look for evidence of ground movement -land slip -subsidence
	Internal corrosion	Magnetic Flux Leakage (MFL) / US / Remote Field Eddy Current (RFEC) pigs	US X-ray MFL
	External corrosion	MFL / US /RFEC pigs	Visual examination shot blast + measure US X-ray MFL

Although some techniques may be duplicated, there may be differences due to material type i.e. what might work on one material and diameter may not be suitable for another. There were a number of outcomes from this exercise:

- A number of known technologies that could be considered as potential joint condition assessment techniques were identified.
- There were several parameters identified where no condition assessment technique was currently available.
- Even where it was considered that a technique was available or could be developed, there was no historic data to compare the limits for that parameter. For example, what is the extent of degradation since time of installation for angular deflection? There may also be certain limits imposed on some joints, which may be taken as a 'joint at risk' in terms of maximum deflection.

This study also raised a number of practical issues that would need to be addressed:

- How does condition relate to potential for failure? e.g. although leakage would be an obvious sign of severe degradation, there is a need to assess how that degradation can be detected before evidence of leakage becomes apparent.
- How, and to what extent, can the degradation modes be used as indicators of condition?
- The original state of the joint is unknown except that the joint was 'sound' at the time of the pressure test.

From this study, it became apparent that many of the permutations presented similar possible degradation modes and that similar assessment techniques would need to be considered. For example, angular deflection is a potential problem for most joint types, irrespective of pipe material, and any technique capable of quantifying deflection would be equally applicable to all joint types influenced by this degradation mechanism. A greater understanding of joint mechanisms is required in order to determine the influencing components of joint integrity. It would then be possible to address the degradation process and influencing loads or condition on those components in terms of long-term performance. To assist with this, some initial finite element analysis work was undertaken on lead yarn joints, since it appears to be the most common joint, and on the limited data available, presents the most common joint problem. This work is described in Chapter 5.

Effect of the Environment

External

The ground surrounding the pipeline can have an influence on its long term integrity. Iron, steel, galvanised iron and copper systems are all vulnerable to attack from the presence of acids, sulphates and chlorides in the soil and depending upon the pH, can quickly become corroded unless they are coated or wrapped. This can apply to the joint bell and spigot as well as bolts and fittings maintaining the joint integrity. Asbestos cement and concrete pipes can be attacked by acids and are sensitive to the presence of sulphates. Plastic pipes are generally unaffected by inorganic substances such as acids or alkalis but organic compounds can cause problems. Although the area presented tends to be very small, the seals used in mechanical joints can be susceptible to swelling and deterioration in organic chemicals found in the surrounding ground. The chemicals may cause embrittlement or stress relaxation in the seal, leading to a loss in performance and potential leakage.

Internal

Specific water quality conditions can enhance the degradative effect on sealing materials within water distribution networks. Elastomeric polymers are subject to oxidative attack from agents within the purveyed water such as chlorine. However, whether exposed to free chlorine, chloroamines or no chlorine at all, elastomers in water distribution systems can have a more limited life span. The nature of the elastomer failure usually involves some form of swelling (due to water absorption) that can grossly distort its shape leading to surface cracking and further degradation.

The introduction of more stringent trihalomethane standards has produced significant changes in the water industry in the United States, including in some cases a change over to chloroamine disinfection as a means of reducing THM production. As chloroamination has been reintroduced, several utilities have reported problems with substantially higher rates of elastomer failures. The implications of chloroamine-related elastomer failure are far reaching due to the threat posed to the numerous gaskets, glands, compression fittings and valve bodies utilised throughout a typical distribution system.

An AwwaRF study was completed in 1993 (Reiber), where accelerated tests showed that solutions of chloroamines (either monochloroamine or dichloroamine) produced greater material swelling, deeper and more dense surface cracking, more rapid loss of elasticity and greater loss of tensile strength than equivalent concentrations of free chlorine. The elastomers most susceptible to attack are those formulated with natural isoprenes (rubber) or synthetic isoprene derivatives. The more recently introduced synthetic polymers performed better. Although the tests were performed upon thin samples immersed in the solution, the results would suggest that joint gaskets would also be susceptible to deterioration over time by this mechanism (depending upon concentration, temperature and stress level). A technique to remove a sample and test the condition of a rubber gasket may therefore prove to be beneficial.

Understanding Joint Performance

The objectives of this aspect were to identify the key parameters influencing:

- Initial joint sealing integrity
- Potential joint degradation mechanisms
- Condition classification in a practical context

Some joints are made up of a combination of systems (as shown in Figure 4.1), ranging from welded metallic or plastic components to multi-component mechanical fittings.

Courtesy Northumbrian Water Ltd.

Figure 4.1 Combination of joint types

A technical search was carried out to establish information relating component construction and performance such as seals, gasket, creep and other physical properties of lead etc. This search covered:

- IMechE papers
- AwwaRF
- UKWIR
- International conference papers
- Manufacturers data direct and Internet
- Library
- Standards

There is a need to better understand the factors influencing joint sealing mechanisms and consideration can then be given to potential failure mechanisms. This will enable a more pro-active, preventative approach rather than reactive approach to be adopted. A sort of

'Cause – Mechanism – Symptom ' approach
e.g. coating failure – corrosion – rust
ground movement – pull out – leak

Hence the need to inspect coating and repair as necessary before corrosion occurs and to monitor ground movement for possible subsequent loss of joint integrity.

JOINT DEGRADATION

It is important to note that, as with pipes and humans, joints age from day one. Initial condition is described as sound at the time of commissioning since they have passed a pressure test. As with humans, each joint will have its own in-built characteristics and will be subjected to different operational, environmental and loading regimes. Each joint is individual and its initial ‘sound’ condition could range from a poor to good predisposition to operational factors and loading regimes. As time passes, a joint’s integrity to maintain a seal will deteriorate until eventually it will leak. That time may be anywhere from day one to perhaps 1000 years. The following parameters are considered to influence joint integrity throughout its service life.

- Initial condition – passes pressure test
- Deterioration – variable rates
- Good installation – expect normal service life
- Poor installation – possible reduced service life
  - Poor quality materials – e.g. composition/processing
  - Incorrect materials or dimensions
  - Poor installation technique/misalignment
  - Equipment malfunction e.g. welding tool

All joints will be exposed to some form of loading regime both internal and external, such as:

- Internal
  - Normal operating pressure
  - Cyclic pressure variations
  - Surge/water hammer
  - Seasonal temperature changes
- External
  - Traffic
  - Overburden
  - Ground movement
  - Corrosion
  - Erosion
  - Soil condition
  - Seasonal temperature changes
  - Water table/ground moisture

External soil/water aggressiveness, especially salt water, can be a probable source of premature seal/gasket failure since natural degradation from internal water is considered to be minimal. Seasonal ground conditions leading to different water levels, frost heave, dry ground can lead to local ground movement, putting additional load on the joint. Joint performance can also be affected by gradual or sudden events, such as;

- Gradual
  - Relaxation/creep of joint materials e.g. gasket/seal/lead
  - Degradation of yarn/gasket etc due to; age/stress/water/microbiological attack
  - Reduction of Bolt Tightness
  - Corrosion – see Figure 4.2
  - Embrittlement/age hardening
  - Integrity of welds

- Sudden
  - Ground movement – slip/seismic- see Figure 4.3
  - Surge

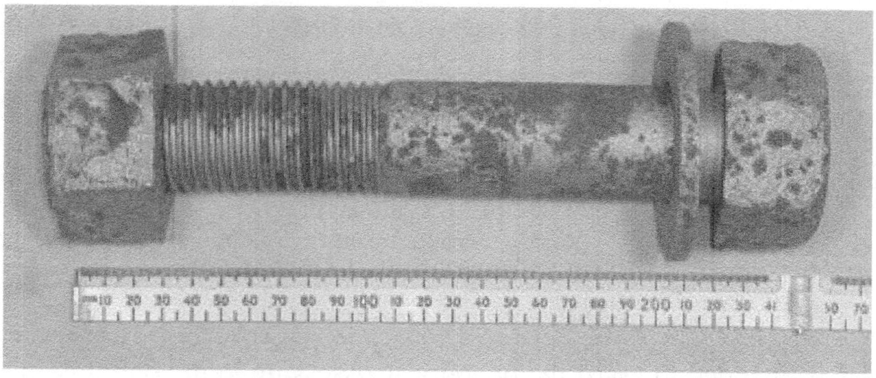

Courtesy Advanced Engineering Solutions Ltd.

Figure 4.2 Example of early stages of gradual bolt corrosion

Courtesy Northumbrian Water Ltd.

Figure 4.3 Example of sudden ground movement – landslip

As mentioned earlier, the extent of any leakage and consequence, is commonly viewed as the accepted condition of the joint. The severity of any detectable leak will determine its classification as 'failure' when an operational response is required. There has been some debate on this subject and the project team decided that a form of degradation ranking based on leak classification (Czernik, 1996) should be prepared to assist in both a qualitative and quantative representation of condition, shown as Table 4.2. The level of detection by currently available techniques is also shown.

Table 4.2
Leak classification

Degradation	Class	Condition	Detectable
None	0	Sound	Exposed & visual only
Minimal	1	Dampness – no droplets	Exposed & visual only
Slight	2	Local Droplets	Exposed & visual only
Minor	3	Droplet bead around circumference	Exposed & visual only
Moderate	4	Drip @ 1 drop/5 mins	Exposed & visual only
Significant	5	Drip @ 1 drop /5sec (0.2 US gals/day)	Possible surface pool
Severe	6	Steady stream (≥20 US gals/day)	Instrument & pool

Data adapted from Czernik 1996.

Table 4.2 is intended to assist the operator in assessing joint condition in a more practical way in relation to operational requirements. However, it should be pointed out that this ranking applies to factory plant pipe work and would not necessarily apply to water utilities assessments of severity of leakage. Nevertheless, it has been used as a basis for ranking joint condition throughout this report.

Only joints in the significant or severe degradation classes are likely to be detectable by currently available instrumentation and leak detection equipment from surface observation. However, where there has been long term minor leakage, there may be some noticeable change in vegetation or ground condition in the dry season by satellite/aerial observations. A practical example of this ranking relating to angular displacement from ground movement, shows minimal degradation with Class 1 leakage (Figures 4.4 and 4.5).

Courtesy Northumbrian Water Ltd.

Figure 4.4 Angular displacement from ground movement

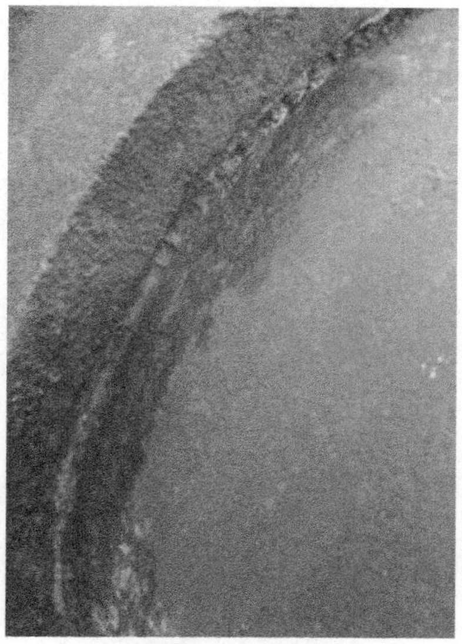

Courtesy Northumbrian Water Ltd.

Figure 4.5 Class 1 Weep and dampness but no droplets

In this example, the pipe is partially exposed and the condition can be readily assessed visually. If buried, this weep would be undetectable. What are the physical parameters now causing this joint to weep and are they measurable? This example also highlights the difficulty in quantifying limit states since some pipe joints are capable of tolerating some angular deflection. Socket and spigot and pipelines are often deliberately laid with angular displacement in order to follow a contour or fall line around a slope.

Initial installation is key to a joint's life but what tolerances are there that might provide an indication of joint integrity. Lead yarn joints are the most common but required skills to install, and therefore are prone to operator variance and possible error. A further example of a pipe joint failure due to rotation of the joint as a result of reduced support is shown on an 8 inch (200 mm) diameter cast iron pipe at Percy Park Metro, where there has been vertical pipe movement but no horizontal movement (Figure 4.6).

Courtesy Northumbrian Water Ltd.

Figure 4.6 Vertical displacement at Percy Park Metro

In this case, the pipe has rotated about the springing level by 7 degrees and as seen in Figure 4.7, there is evidence of some dampness at this position after repair, but none at the crown or invert.

Courtesy Northumbrian Water Ltd.

Figure 4.7 Discolouration at springing level

Generally, more catastrophic pipeline failures relate to structural failure of pipe barrels. Structural joint failures are typically rare on large diameter pipes but where they occur the consequence would be considered catastrophic. However, where there is a loss of joint integrity in the form of minor failure due to a seal problem for example, the consequence would be seen as non-urgent. However, prolonged leaking from a joint or pipe barrel pinhole leak can result in washout of bedding and could result in loss of support and in consequence, a catastrophic joint or barrel failure. This would lead to:

- Loss of supply
- Loss of product
- Threat to the environment, compensation claims etc.
- Immediate repair necessary

Thus it must be understood that whilst minor leakage may not be viewed as problematic, it could lead to subsequent operational failure. The flowchart, Figure 4.8, shows all the influencing parameters in respect of joint condition throughout its life cycle.

In summary, there are many factors that influence a joint's initial and long term performance. An awareness of these factors, along with experience of local conditions and operational problems will assist water company operatives to consider how best to manage their assets.

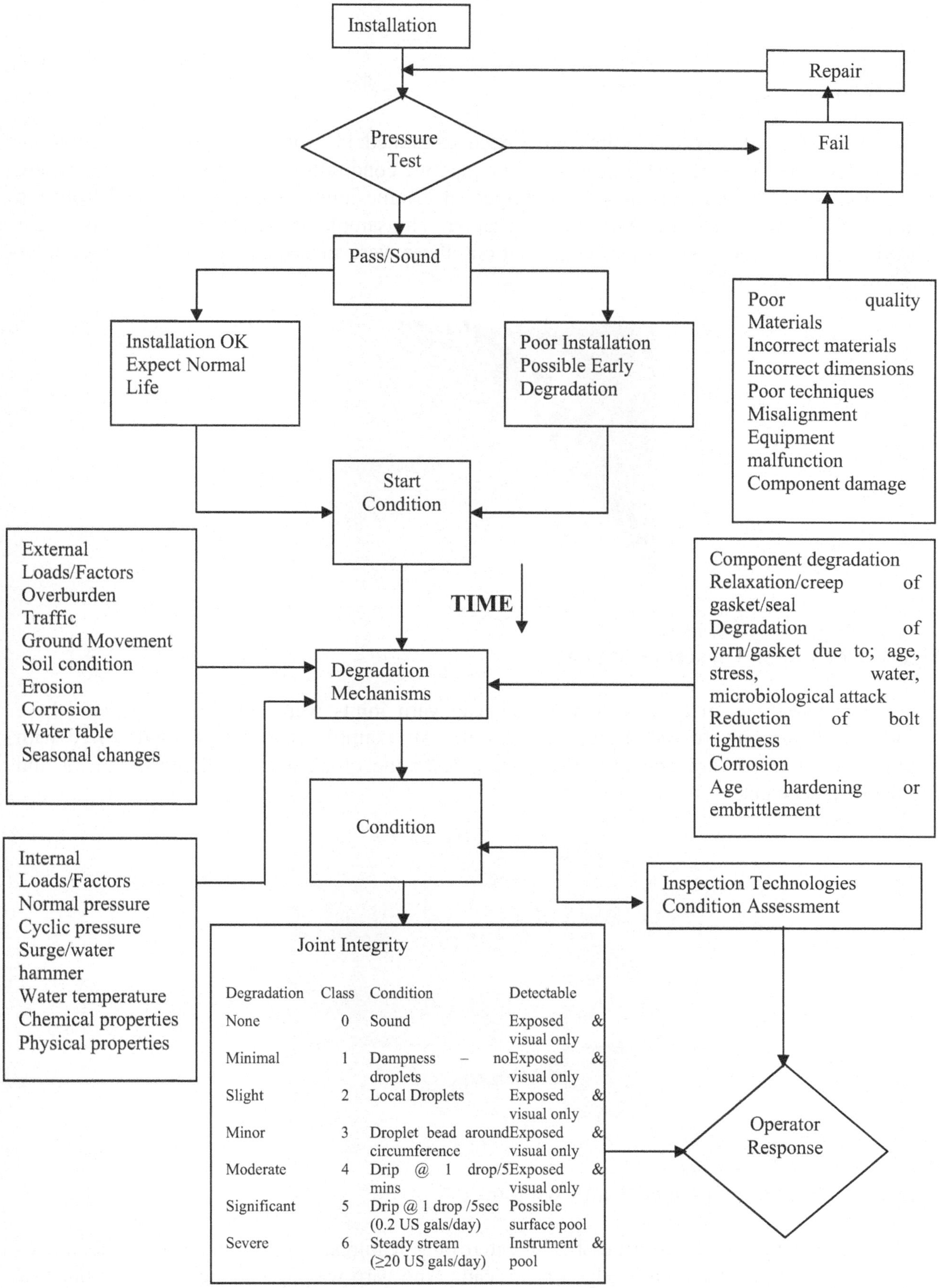

Figure 4.8 Parameters influencing joint condition

Joint Samples

A number of sectioned joints have been examined in order to assess their condition. These joints were randomly selected as part of a pipeline condition assessment programme where the section of the pipeline is cut out and inspected. Some joints were also provided from the failure analysis work described later in this chapter. The samples were cut in half and then grit blasted. Where a joint exists as part of the cut out, these were subsequently inspected. Examples are shown in Figure 4.9.

Courtesy of Northumbrian Water Ltd.
Figure 4.9 Examples of sectioned joints

The variations in construction of the lead yarn joints were significant. Despite these variations, all joints were sound prior to cut out. An example is shown in Figure 4.10 with different amounts of yarn and lead from one side to the other, and small misalignment and angular displacement.

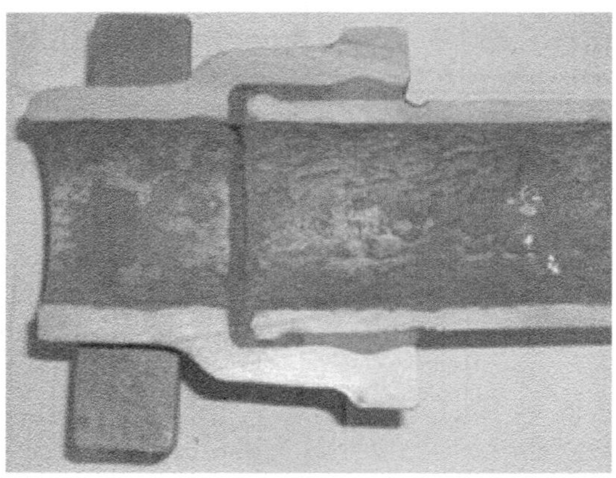

Courtesy Northumbrian Water Ltd.
Figure 4.10 Section of a typical lead/yarn joint

This example revealed a number of interesting features. Firstly, there was evidence of pipe barrel corrosion at the back edge of the lead. The yarn was in a sound condition but was starting to become friable, probably due to drying out. Also, it had been installed effectively as a spiral since the extent of the yarn on the bottom went just beyond the spigot whilst on the top it went further past the end of the spigot. Consequently, the amount of caulked lead was variable. The joint gap was to within British Standard limits but there was some evidence of misalignment. Nevertheless, the joint was sound but due to the corrosion, there would be a weakness around the circumference at the lead extremity if the pipe was subjected to bending. A number of joints with an internal socket depth of 3 inch (75mm) were examined to compare dimensions detailed in Figure 4.11 and in Table 4.3.

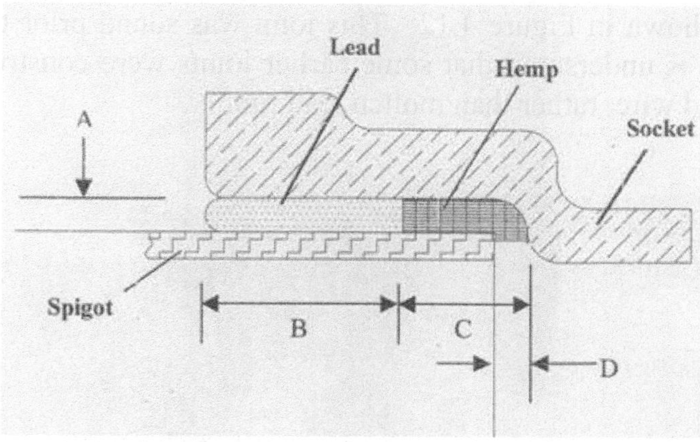

Courtesy Advanced Engineering Solutions Ltd.

Figure 4.11 Lead/yarn joint dimension reference

Table 4.3
Lead/yarn joint dimensions (mm)

Sample No./Side	A	B	C	D
1 RH side	8.2	42.5	10.3	4.3
1 LH side	7.6	41.4	11.3	2.7
2 RH side	8.9	50.0	24.7	2.0
2 LH side	5.6	53.3	22.2	1.8
3 RH side	8.7	44.0	28.7	1.8
3 LH side	6.8	58.7	16.6	2.5
4 RH side	6.3	45.8	30.8	3.5
4 LH side	8	54.6	20.2	4.2
5 RH side	7.4	33.1	42.7	6.7
5 LH side	5.7	46.5	29.8	7.9

The results show:

- Misalignment or angular displacement 5.6-8.9 mm (A)
- Different amount of lead: 33-46 mm (B) on one pipe: 33-58 mm range
- Different depth of yarn: 42-30 mm (C) on one pipe: 10-42 mm range
- Joint Gap: 2.7-4.3 mm (D) on one pipe: 1.8-8 mm range

The inference from these measurements is that if inspection tools/techniques could be developed to assess the amount of yarn or lead present or measure misalignment, it would still be difficult to assess condition from this since the joints still appear to remain sound over a wide range of tolerances. However, measuring joint gap and angular displacement does merit further consideration since if regularly monitored, changes in pull out or angular displacement following ground movement could be detected.

There are even more extreme examples such as little or no yarn present such that the lead runs into the pipe as shown in Figure 4.12. This joint was sound prior to cut out showing that lead alone will seal. It is understood that some earlier joints were constructed without yarn and using caulked solid lead wire, rather than molten lead run.

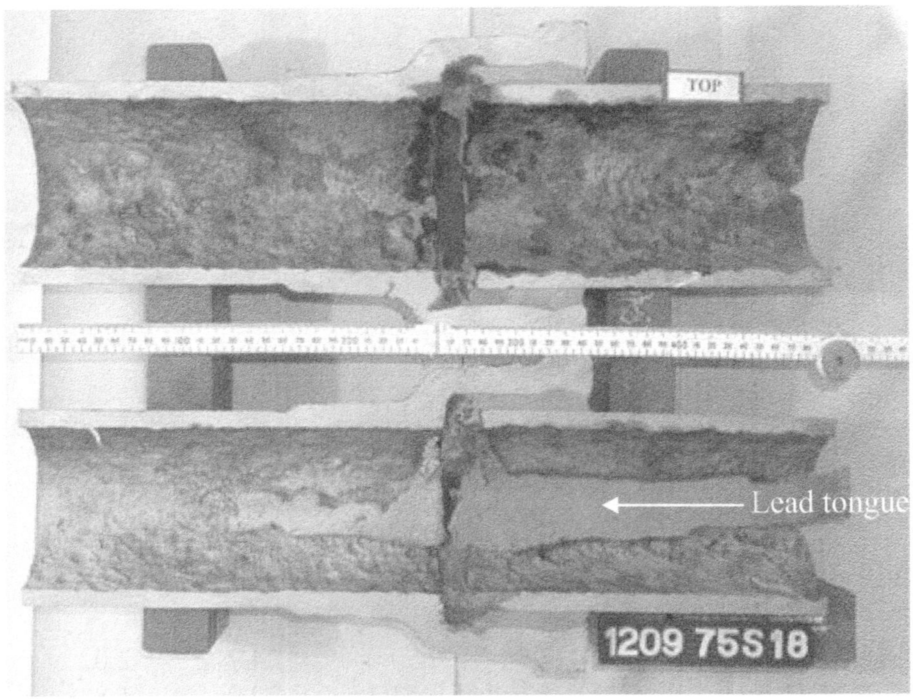

Courtesy Northumbrian Water Ltd.

Figure 4.12 Lead tongue in cast iron pipe

Since it is unclear whether there is shear failure of the lead or loss of bonding between the lead and pipe surface in the event of pull out, then the effect of gradual temperature change between summer and winter extremes could be considered. The actual physical movement taken up on a 12 feet (3.65 m) long cast iron pipe over a 20°C winter to summer temperature difference at pipe depth is 0.033 inches (0.85 mm). Whether this effect is cumulative over a number of joints or taken up by each joint is yet another unknown factor. Where pipes are above ground, then the daily temperature variations may have some effect on the expansion and contraction. However, it is considered that the prevailing water temperature, which will fluctuate very little daily, and the heat transfer coefficient of flowing water will dominate. Although there will be solar gain, the net effect is considered that there will be very little daily expansion and subsequent contraction however, it is recognised as another possible factor.

Examination of a number of elastomeric seal socket and spigot joints showed no apparent degradation of the seal. Flexibility and hardness seemed similar to new seals. What may be of more interest is the corrosion on the spigot and socket found upstream of the seal on one example.

In summary, socket and spigot joints are clearly capable of performing satisfactorily with a wide variety of lead and yarn content, joint gap and initial angular displacement. It therefore seems impractical to develop techniques to measure these parameters since they provide no indication of condition or projected life before leaking. Monitoring change is unlikely to provide performance data except in the case of angular movement or pull-out in areas prone to ground movement.

GASKET RIG

In order to try to better understand the requirements for effective sealing, a number of different gaskets were prepared and located between two coarse metal plates. The range of gaskets included different material types, wall thickness and cross sectional area in order to evaluate any differences in performance and hence the requirements to form an effective seal.

Theory

Many flanged pipe joints are present in water distribution networks, with rubber gaskets as seals. Gasket seal technology is complex and covers many different product types and working environments. Sealing is not a precise science and much of the available data comes from experimental work. Even today, investigations into the properties of rubber gaskets are currently under way in relation to their long-term characteristics under compression. Initial seating loads are needed to activate a seal, irrespective of line pressure and then secondary loads to maintain that seal. Reference has been made to a number of seal technology books, (Buchter and Czernik) with the latter describing many of the design problems that a purpose made gasket would need to address. There are a number of relevant American Society of Mechanical Engineers (ASME) codes for gaskets relating specifically to:

- Fluid type
- Temperature variations
- Flange surface condition
- Seating pressures

It must be noted that rubber tends to be displaced under load due to shear forces and essentially does not compress. The characteristics and long term performance of a rubber gasket are therefore critical with regard to its effectiveness in a specific working environment. Consideration needs to be given to the following parameters:

- Macroconformability and Microconformability – that is the materials ability to deform and flow into the coarse and fine contours of the surface
- Bulk modulus, ASTM D412
- Hardness, ASTM D676, D314, D531
- Resilience
- Hysteresis
- Modulus of elasticity, ASTM D575, D79, D945

The initial seating load, described as the primary bolt load, allows material to bed/flow into flange face and is expressed as the minimum seating pressure required to create a seal. The load depends upon gasket type, composition and resilience. For example, the seating load is 200 psi for plain rubber and 6000 psi for PTFE. The primary bolt load W1 is calculated from:

- Primary bolt load W1, is calculated from:

  W1 = πbGy

  Where b = effective gasket contact width

  G = mean diameter, depending upon gasket width

  y = required seating pressure for that gasket – typically determined by experiment

- Secondary bolt load W2, which is required to maintain the seal against line pressure and is additive to the primary bolt load, is calculated from:

  W2 = πG2P/4 + 2πbGmP

  where P = Maximum working pressure

  m = gasket factor (depends upon type, thickness etc. and is typically 0.5-6.5)
- Total bolt load =W1 + W2

Long term performance of a gasket depends upon:

- Stress Relaxation, change in stress at constant strain
- Creep (compression drift), change in strain at constant stress
- Combined effect termed 'Creep Relaxation' referred to as relaxation
  - Loglog function
  - Thicker gasket - more relaxation
  - Higher bolt load - greater density - less relaxation

The above factors will influence the initial gasket and joint condition, and these need to be considered for their sensitivity and contribution to joint integrity. A greater understanding of the initial and degrading factors that influence joint performance assists with the identification of possible techniques to measure those factors.

Some limited experimental work was carried out by AESL to investigate the use of this gasket theory in practice and to gain a better understanding of rubber performance as a sealing medium for joints.

Gasket Performance Tests

A variety of potential gasket materials and designs were tested using a specially constructed test rig and the sealing performance measured over a range of simulated operating conditions. The gasket test rig is similar in operating principle to that used for ASME tests and comprises a 10 ton capacity press to simulate the applied pressure on a gasket for a flanged joint and for live joint inspection and rehabilitation techniques. The rig is shown in schematic form, Figure 4.13, and operationally, Figure 4.14.

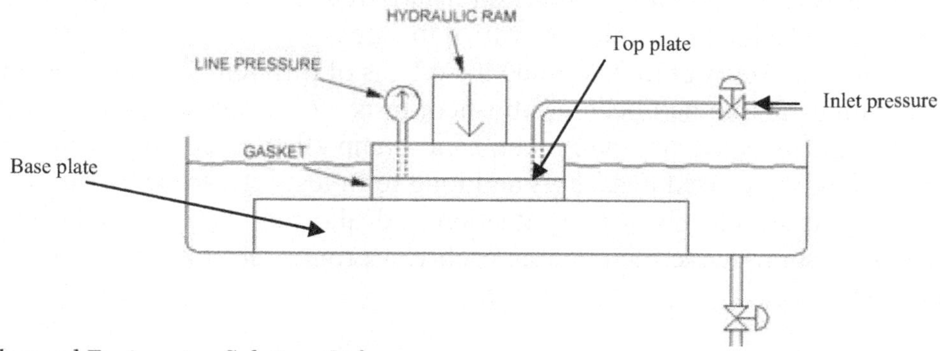

Courtesy Advanced Engineering Solutions Ltd.
Figure 4.13 Schematic of gasket test rig

Courtesy Advanced Engineering Solutions Ltd.
Figure 4.14 Gasket test rig

The gasket contact area is measured for each sample from which the interfacial contact pressure can be calculated. The rig was calibrated using a load cell thereby providing an exact measure of interfacial pressure. The gaskets were prepared with a cut out section as shown in Figure 4.15. he gasket is sandwiched between the top and bottom plate with an applied load. The cut out centre section of the gasket is then pressurised with water. The applied load and water line pressure are monitored. All leaks were noted with a loss of pressure.

Courtesy Advanced Engineering Solutions Ltd.
Figure 4.15 Waffle gasket with cut out centre section

Gasket performance tests were conducted at different simulated line pressures up to 16 bar where practicable. The loading cycle was varied to assess any change in performance.

Tests have been carried out to determine the effects of different loading procedures. Once the necessary operating regime had been established, tests were carried out using either a hand pump or full 10 bar mains water pressure. The hand pump allowed control of the pressure and facilitated different loading procedures, short and long term tests etc. to be assessed. The mains water tests simulated more closely a real situation and allowed 'leak' tests to be conducted. Whilst detailed observations were recorded as each test progressed, the objective of most tests was to determine;

- The level of 'seating' pressure required
- The interfacial pressures needed to seal

Different gasket materials, thickness, section etc. as listed below, were prepared and sandwiched between two coarse 'flat' plates, to measure the effective seal.

- Waffle gasket 'broad'
- Waffle gasket 'narrow'
- Flat thin
- Flat thick
- Ribbed thin
- Ribbed thick
- V belt

The waffle, ribbed and flat gaskets were bonded to the top plate using Loctite Sealastic 578 in order to prevent leakage from the underside of the test specimen. Whilst this worked satisfactorily, on some tests the lateral loads across the section caused extrusion of the material. For the V-belt test, where there is only point contact, a support plate was prepared as shown in Figure 4.16.

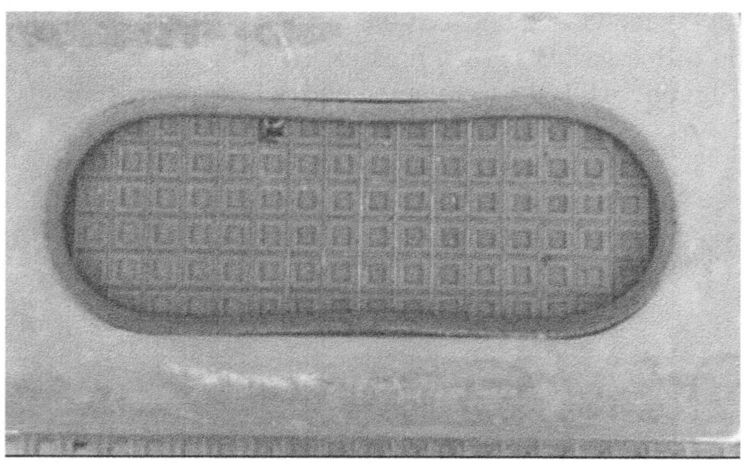

Courtesy Advanced Engineering Solutions Ltd.

Figure 4.16 V belt support plate

Thin plates were also prepared to simulate possible different pipe defect patterns, as follows;

- Chevron slots
- 1.5mm longitudinal 'V' groove
- 2mm longitudinal 'V' groove
- Weld splatter
- 2mm round nosed groove

In Figure 4.17 the chevron slot defects simulate damage to the seal during installation, or raised areas on the external surface of the pipe which may prevent sealing contact.

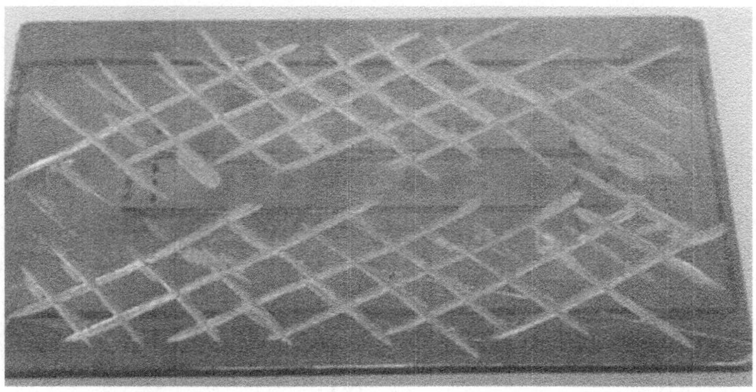

Courtesy Advanced Engineering Solutions Ltd.

Figure 4.17 Chevron slot defect simulation

In order to simulate adverse pipe conditions, tubercule growth deposits were scraped from a cut out section of pipe and sandwiched between the gasket and top plate. It is important to note that the design of the rig is akin to a flanged joint gasket. Consequently, when the internal pressure is increased then the resulting force on the underside of the exposed top plate will, and does, push the plate up against the applied down load. This has been allowed for in the results. Dial gauge readings were recorded in order to assess plate movement and thereby measure the amount of gasket compression. Long-term tests showed a degree of weeping back past the hydraulic piston seals, resulting in some loss of applied pressure.

Gasket Performance Test Results

There were a large number of tests carried out in order to gain data on the performance characteristics of the different gasket materials, the loading procedure etc. These have been invaluable in establishing a practical understanding of the sealing process. The results detailed in Tables 4.4, 4.5, 4.6 and 4.7, summarise the major observations. Table 4.4 shows the results of the required pressure to seat the gasket against 1 bar water pressure (equivalent to W1) and the total pressure needed to maintain a seal against 16 bar water pressure (equivalent to W1+W2). The calculated combined interfacial contact pressure, W2, is also shown. Table 4.5 shows the results of water pressure reached prior to leaking with the narrow waffle gasket with an applied constant interfacial pressure with different defects. Table 4.6 shows the effects of resultant leak rates against 10 bar mains pressure with different amounts of interfacial pressure with different gaskets. Table 4.7 shows the effects of resultant leak rates against 10 bar mains pressure with different amounts of interfacial pressure with different defect using the narrow waffle gasket.

Table 4.4
Seating and seal contact pressure for different gaskets at 16 bar water pressure

Rubber Gasket Type	Primary seating pressure (psi) W1	Secondary seating pressure (psi) W1+W2	Interfacial contact pressure (psi) W2
Broad Waffle (40mm)	50	230 (16 bar)	180
Narrow Waffle (20mm)	100	140	250
Flat Plain 3mm	80	140	170
Flat Plain 1.6mm	80	220	220
Broad Ribbed	110	100	200
V belt	200	150	450

Table 4.5
Leak pressure with narrow Waffle gasket with different defects

Defect	Water pressure prior to leak (psi)	Interfacial contact pressure (psi)
Chevron	150	190
1.5 mm V groove	80	190
2mm V groove	0	Leak
Weld splatter	120	190
Pipe growth	40	190
Pipe growth flushed	140*	160**

Note: * Water leaked adjacent to the crushed particles
** Gasket extruded due to distortion from crushed particles.

Table 4.6
10 bar mains pressure leak rate tests – different gaskets and interfacial pressure

Gasket	Interfacial pressure (psi)	Leak rate (gal/hour)
Flat Rubber 3mm	120	13.5
	140	4
	150	1.6
	170	0.8
	180	0
Narrow Waffle	110	33
	120	4.4
	130	0.3
	140	0
V belt	350	90
	390	39
	450	0

Table 4.7
10 bar mains pressure leak rate tests – Waffle gasket with different defects

Defect	Interfacial pressure (psi)	Leak rate gal/hour
Chevron	120	36
	130	26
	140	7.5
	150	3.2
	160	1.6
	170	1.1
	180	0
Weld Splatter	110	36
	120	10
	130	3.8
	140	1.8
	150	0
Pipe growth – partial flush	140	35
	160	10.8
	190	0

Findings

It can be seen from Table 4.4 that the different thickness plain rubber gaskets performed in a similar manner with seating pressure ranging from 50 to 110 psi depending upon shape and thickness. The reinforced rubber V belt required a much higher seating pressure of 200 psi due to its stiffness. Since the unsupported gasket material tended to extrude, the water pressure had to be limited on some tests. Test results show that the interfacial contact pressure to seal up to 16 bar ranged from 180 psi for the waffle gasket to 220 psi for the flat rubber gasket. The reinforced rubber V belt, which effectively acts as a single point contact, required 450 psi to seal at 10 bar.

These results clearly show that a satisfactory seal can be achieved against 16 bar line pressure at around 200 psi interfacial pressure using just standard gasket material. This is similar in magnitude to the value for rubber described in the experimental section.

Since the waffle gasket, performed marginally better than the ribbed and flat gaskets and is designed specifically for external wrap round pipe repair collars, this material was used for the defect tests, detailed in Tables 4.5 and 4.7. The results show that a complete seal was achieved with the chevron up to 150 psi (10 bar), and weld splatter defects up to 120 psi (8 bar). No seal was achieved on the 2mm deep V groove and only a partial seal on the 1.5 mm deep groove. These defects are very severe and do not necessarily represent pipe defect conditions but the tests do show the limitations of the gasket under such conditions.

The pipe growth tubercules used for the tests were typically about 30 mm high and crushed easily when a load was applied. The crushed particles however tended to compact down to a 3-5 mm high spot causing local deformation of the gasket. Attempts were made to flush the particles as they were crushed and this was relatively successful, Figure 4.18. Since a positive flushing pressure could not be created whilst there was no effective seal the extent of flushing was limited. In a live situation, it is anticipated that positive flow would remove most of the particles and any residual deposit could be accommodated by the waffle contour.

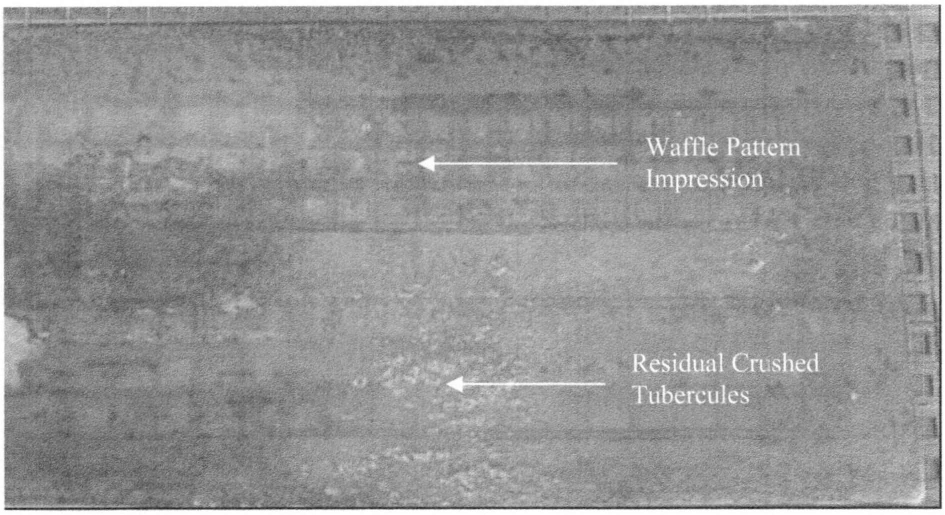

Figure 4.18 Residual crushed tubercules

The results from the leakage tests at 10 bar with reduced contact pressures, Tables 4.6 and 4.7, have provided data on 'partial seals'. The consequence of loss of interfacial pressure due to relaxation of the gasket material, bolt tightness etc. is quite evident.

Gasket Performance Tests Conclusions -Bolted Flanges

The tests have confirmed that good gasket performance depends upon appropriately designed gasket, clean faces and correct level of torque to provide the necessary seating and sealing forces. Once installed, there is no readily accessible parameter that could be measured to provide an indication of joint condition, although the rubber will degrade over time. However, on full faced flanges, the external ultrasonic technique described in Chapter 7 may have the potential to assess the overall soundness of the joint. The only other accessible components are the bolts and evidence of any corrosion.

DUCTILE IRON PIPE BEND TESTS

In order to understand the possible limit state conditions of socket and spigot joints with an elastomeric seal, a test was set up to record the strain induced in the pipe relative to applied bending loads and resultant angular deflection.

An DN 80 mm ductile iron pipe to BS EN545 was used with minimum and maximum spigot insertion depths as recommended by Saint Gobain Pipelines. Schematics of the section of the joint with different insertion depths are shown in Figures 4.19 and 4.20.

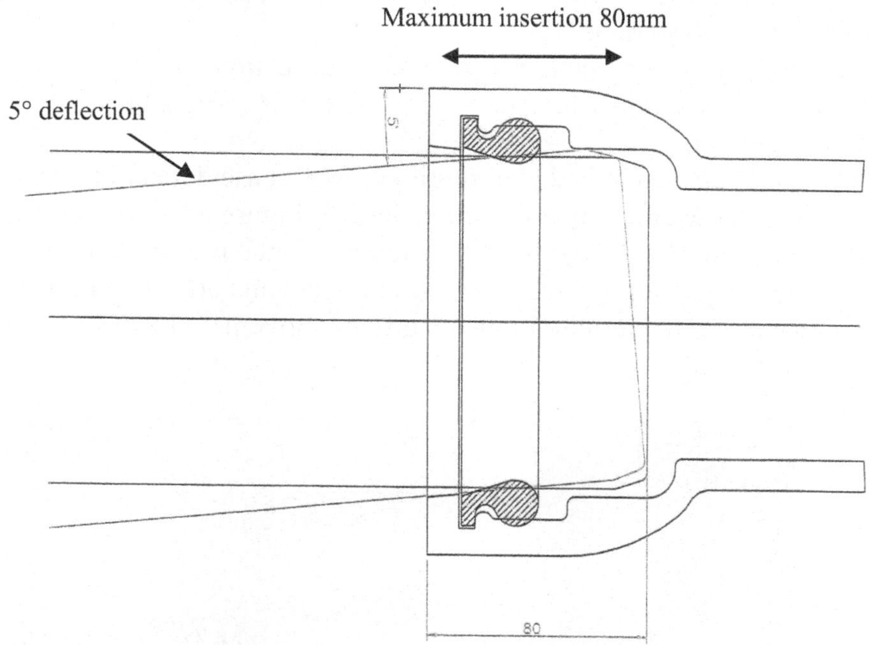

Courtesy Advanced Engineering Solutions Ltd.
Figure 4.19 Limit state angular deflection ~ maximum insertion depth

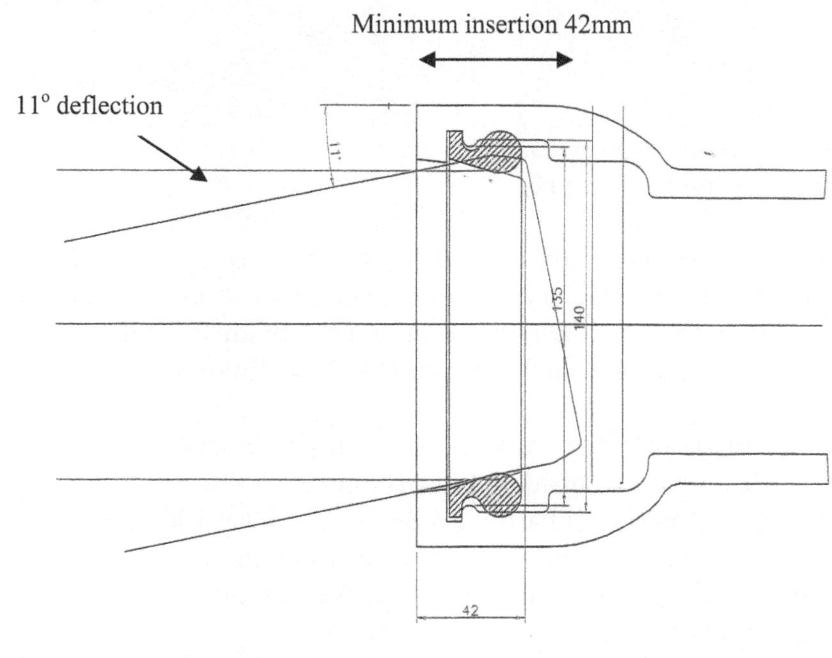

Courtesy Advanced Engineering Solutions Ltd.
Figure 4.20 Limit state angular deflection ~ minimum insertion depth

These show that angular deflection to reach absolute limit state (i.e. become metal bound) is dependent upon the insertion depth 11 degrees for minimum insertion of 42 mm and 5 degrees for maximum insertion of 80 mm.

Pipes are often laid at predetermined angular deflection within the limits of manufacturers recommendations. The extent of insertion therefore needs to be adjusted relative to the amount of deflection to be applied.

The test rig set up was intended to represent worst case deflection where the pipe is totally unsupported at the joint over a nominal 2 metre length, Figure 4.21. This situation could occur if there was wash out of local side and bedding support. Since it was not known what the resultant strain might be, loads were applied in 20 lbs (9kg) increments up to a maximum of 100 lbs (45kg) for safety. It was not the intention to replicate ground movement loads but simply to determine if a steady state was reached.

Courtesy Advanced Engineering Solutions Ltd.

Figure 4.21 Ductile pipe bending rig

Standard foil resistance strain gauges (RSG), were attached at positions of expected highest stress. These gauges were fixed with conventional strain gauge cement and paint. It was not possible to position them on the radius section nor obviously at the point of pipe contact. The positions of the strain gauges are shown in Figure 4.22 as follows:

1 circumferential on socket crown outside – inset 10 mm
2 axial on socket crown outside – inset 10 mm
3 circumferential on socket barrel inside invert – inset 100 mm
4 axial on socket barrel inside invert – inset 100 mm
5 circumferential on spigot crown inside - inset 15 mm
6 axial on spigot invert inside - inset 15 mm
7 circumferential on spigot at crown inside - inset 70 mm
8 axial on spigot at crown inside - inset 70mm

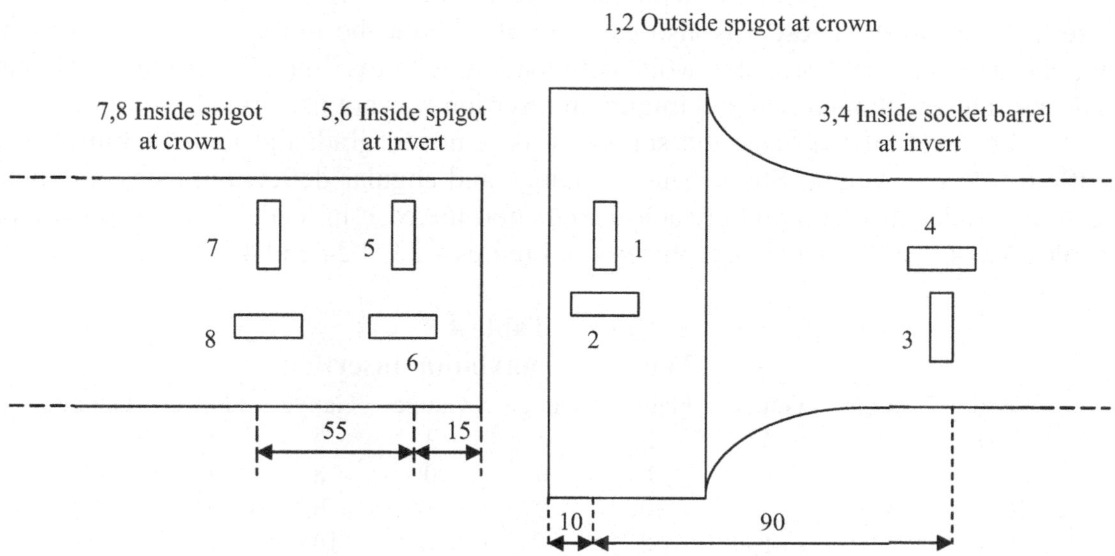

Plan View - mm

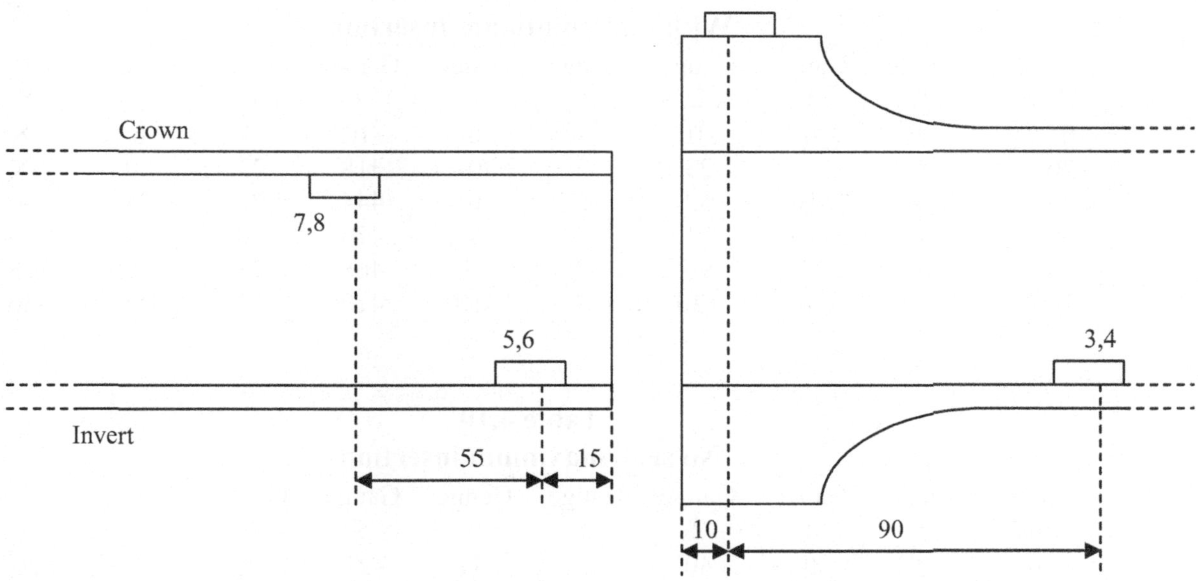

Side View

Figure 4.22 Strain gauge positions (mm - not to scale)

The strain gauges were set up as a ¼ Wheatstone Bridge with a 10 volt supply. Each gauge was balanced at 1000 microstrain (μstrain) as the datum, with the pipe fully supported in a horizontal position. All the readings presented in the tables and graphs represent the tensile (positive) and compressive (negative) changes in μstrain from that datum.

Tests were carried out with the pipe at minimum insertion (42 mm) and maximum insertion (80 mm). A test was also carried out without the rubber seal to determine maximum angular deflection and probable worst case loading with maximum insertion. Note that for safety reasons, a test without a seal and minimum insertion was not performed.

For each of the tests, the support was removed gradually and the pipe was allowed to settle under self weight. Strain gauge readings and angular deflection, using a clinometer, were recorded under this load and at each incremental increase in load. The results are presented in Tables 4.8, 4.9, and 4.10 and graphically in Figures 4.23, 4.24 and 4.25.

Table 4.8
With seal, maximum insertion

Load (lbs)	Angle °	Gauge 1	Gauge 2	Gauge 3	Gauge 4	Gauge 5	Gauge 6	Gauge 7	Gauge 8
0	1	0	0	0	0	8	0	-2	0
20	1.5	-5	30	0	-4	76	-15	-10	0
40	2	-14	78	-2	-12	148	-32	-13	7
60	2	-17	80	-2	-24	68	-36	-13	10
80	2	-16	88	0	-34	-2	-38	-12	12
100	2	-20	96	2	-48	-78	-40	-10	14

Table 4.9
With seal, minimum insertion

Load (lbs)	Angle °	Gauge 1	Gauge 2	Gauge 3	Gauge 4	Gauge 5	Gauge 6	Gauge 7	Gauge 8
0	3	0	-10	0	0	-10	0	6	-8
20	3	-6	23	-2	0	318	23	50	-8
40	4	-10	52	-2	0	444	28	74	-4
60	4	-10	71	-2	0	470	30	100	-16
80	4	-7	96	-2	-4	468	28	122	-28
100	4	-8	130	-4	-10	410	23	140	-36

Table 4.10
No seal, maximum insertion

Load (lbs)	Angle °	Gauge 1	Gauge 2	Gauge 3	Gauge 4	Gauge 5	Gauge 6	Gauge 7	Gauge 8
0	2	0	30	2	-11	-21	-22	-8	20
20	2	-2	52	0	-20	-48	-28	-12	34
40	2.5	-4	70	2	-30	-74	-38	-16	44
60	3	-4	90	2	-40	-102	-38	-24	50
80	3	-8	120	4	-50	-134	-38	-28	62
100	3	-8	132	4	-58	-166	-36	-36	72

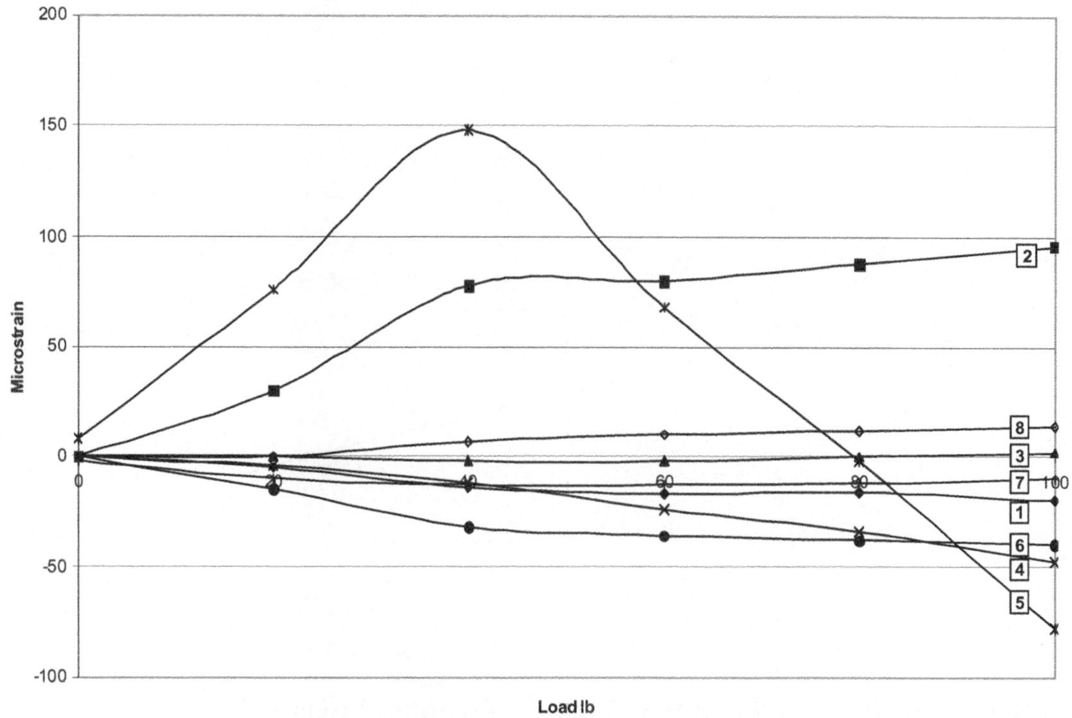

Figure 4.23 Strain ~ Load (with a seal – maximum insertion)

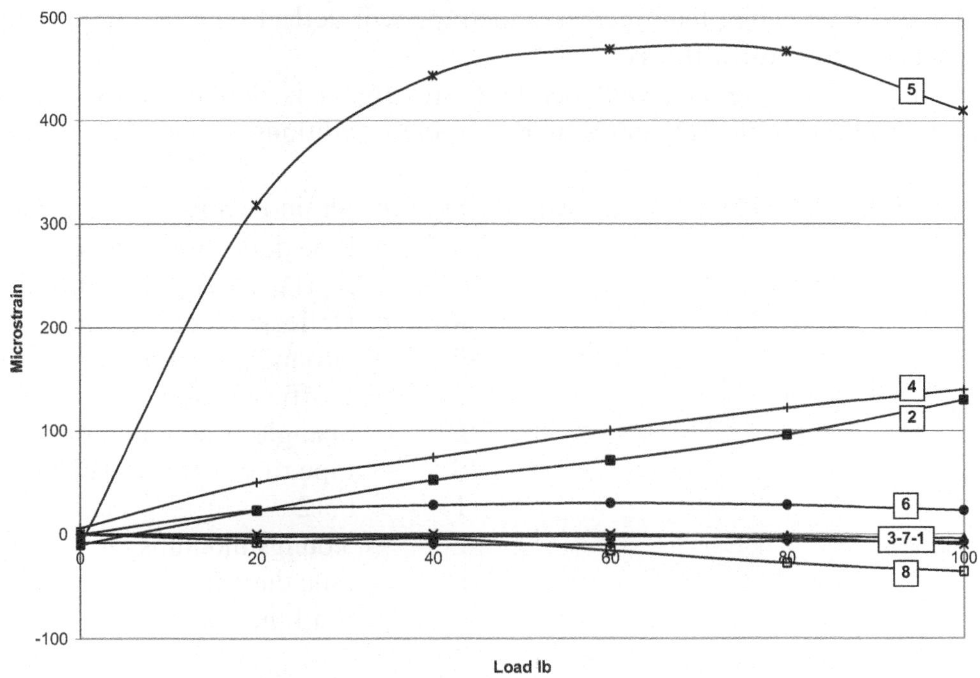

Figure 4.24 Strain ~ Load (with a seal – minimum insertion)

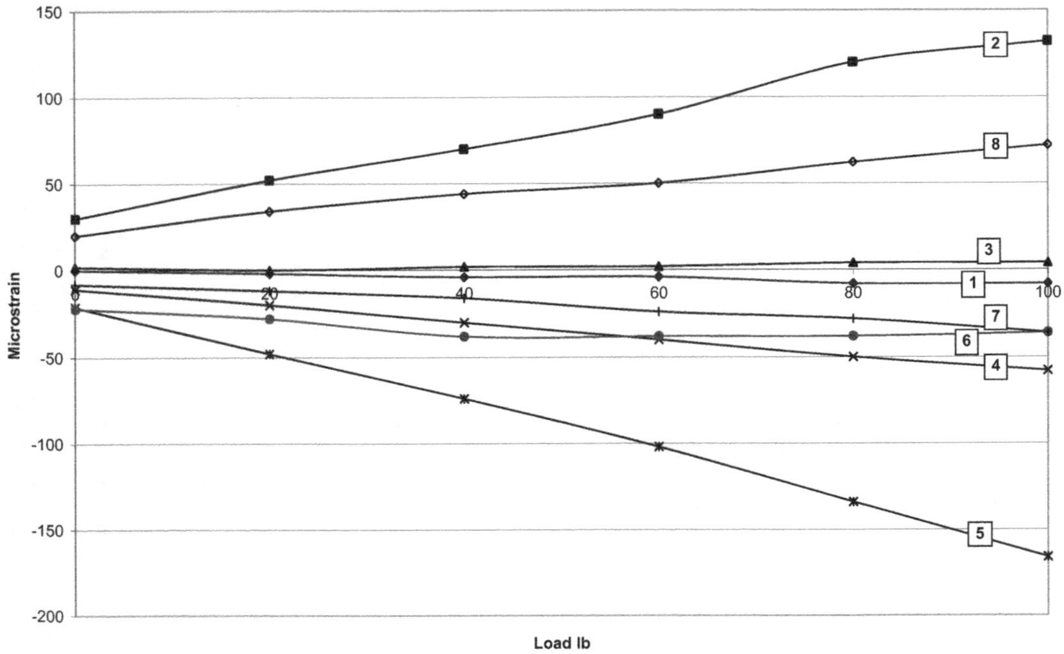

Figure 4.25 Strain ~ Load (without a seal – maximum insertion)

The resultant stresses around the spigot end and at the seal position are quite complex and it is not the intention to attempt to account for all the resultant stresses but to consider the significant outputs. It should be noted that, similar to a ductile iron pipe barrel in soil, as the overburden/traffic crushing load increases the pipe will deflect into an elliptical form depending upon the side support soil stiffness.

The formula for crushing/pipe wall bending stress(σ_{bs}) is defined in BS EN 1295-1: 1998, "Structural design of buried pipelines under various conditions of loading", for semi-rigid pipes as;

$\sigma_{bs} = E\,D_f\,(\Delta/D)\,(t/D)$ where

D_f	=	Strain Factor
Δ/D	=	Pipe deflection/ovalisation
	=	$K_x\,(P_G + P_T)/(8S + 0.061E')$
K_x	=	Deflection coefficient
P_G	=	Ground pressure
P_T	=	Traffic pressure
E'	=	Spangler's soil modulus
S	=	Pipe ring stiffness (EI/D^3)
I	=	$t^3/12$
E	=	Youngs modulus
D	=	Pipe diameter
t	=	Wall thickness

Note that traffic pressure and internal pressure are ignored in this case.

As the pipe deflects vertically, the resistance to horizontal deflection will increase due to the increasing local soil stiffness. At some point, further horizontal deflection will be restricted, and the pipe will tend to stiffen up (effectively try to re-round itself) resulting in redistribution of local stresses into the 45 degree positions. The same will apply at a joint with a seal, and although more complex, the principle is the same. Also it should be borne in mind that for a pipe barrel the side support is assumed to be applied along the length of the barrel whereas for a joint in this test condition the side restraint from spigot deflection is only applied through the seal to the socket.

In the test configuration, as the load is applied, the pipe will deflect both inwards vertically at the crown/invert and outwards horizontally at the springing level thereby deforming the seal locally against the socket support at these positions. At some point, as the seal resistance to horizontal deflection increases then the reaction to increased loads will effectively cause the pipe to try to re-round itself, resulting in redistribution of local stresses, into the 45° positions. At this load, the joint effectively becomes seal bound, with no further significant angular deflection. As the resistance increases at the 45° positions with increased load then they in turn will act as restraints/supports thereby reducing the tensile strain at the crown /invert, eventually developing into compressive strain. Ultimately, with increased loads, the pipe will probably fracture in circumferential bending adjacent to the joint.

The re-rounding effect on the DN80 pipe spigot can be clearly seen from the results of Strain gauge 5, 15 mm inset on the circumferential bore of the spigot at the invert. In the maximum insertion test, the tensile strain increases significantly with load up to 150 μstrain, equivalent to a stress of 1.6T/in^2 (25.5 MPa), with a 40 lbs (18kg) load. At the same load, maximum angular deflection of 2 degrees has also been reached. At this point, full compression of the seal has been reached at the springing level. Thereafter, additional loads result in a redistribution of the strain, as described above, reducing the tensile strain at the invert, which eventually becomes compressive as the spigot becomes re-rounded. Obviously this occurs on a micro scale.

On the minimum insertion test similar results are found but the strain is higher at 470 μstrain, equivalent to a stress of 5.1T/in^2 (80 MPa), with a 60 lbs (27kg) load. This is attributed to the closer proximity of the gauge and probably also to end effects. At the same load, maximum angular deflection of 4 degrees has been reached. Thereafter, the tensile strain reduces and would have become compressive if greater loads could have been applied. The resultant crushing stress of 5.1 T/in^2 (80 MPa) is about 20% of the UTS (Ultimate Tensile Stress) of 26T/in^2 (415 MPa) and the spigot is unlikely to fail. The reason for the larger angular deflection is due to the socket shape for the seal as shown in Figure 4.24. In both these tests, the spigot did not become metal bound.

Where there is no seal, the pipe is effectively free to deflect at the springing level since there is no side restraint due to side clearance. Hence a tensile strain should exist at the crown/invert. However, a compressive strain was unexpectedly registered increasing linearly up to 166 μstrain (see Figure 4.25), equivalent to a compressive stress of 1.8T/in^2 (28 MPa), with a 100 lbs (45kg) load. Close examination of the socket bore in the area of invert contact, showed an irregular shape with two rough-cast protrusions at about the 45° positions. This explains the compressive strain recorded. Nevertheless, the resultant strain is of a similar magnitude to that reached with the seal with maximum insertion. This would suggest that local induced stresses are similar with or without a seal so that if the joint became metal bound then there is no apparent greater risk to failure under the prevailing conditions.

The reduced angular deflection is due to the internal configuration and the slight change with increased load is attributed to small axial slippage since there was no seal friction in this case.

Strain gauge 2, axially at crown of socket, inset 10 mm, shows similar strain on all tests as a reaction to the angular deflection. It increases more or less linearly with load up to 130 tensile μstrain, equivalent to a stress of 1.4T/in^2 (22 MPa), with a 100 lbs (45kg) load.

Strain gauge 7, circumferential at crown of spigot, inset 70 mm, shows similar strain to gauge 2 on the minimum insertion test. It increases linearly with load up to 140 tensile μstrain, equivalent to a stress of 1.4T/in^2 (24 MPa), with a 100 lbs (45kg) load. On the other tests, the gauge registered small compressive strain. The reason for this is unclear but is assumed to be due to its position under different loading regimes.

Strain gauge 8, axial at crown of spigot, inset 70 mm, increased linearly with load up to 72 tensile μstrain, equivalent to a stress of 0.75T/in^2 (12 MPa), with a 100 lbs (45kg) load with no seal. On the other tests, supported by the seal or socket in maximum insertion and or no seal tests, the gauge registered low tensile with maximum insertion and 36 compressive μstrain, equivalent to a stress of 0.4T/in^2 (6 MPa), with a 100 lbs (45kg) load, with minimum insertion. Again, the reason for this is unclear but is assumed to be due its position under different loading regimes. Strain gauge 4, 100 mm inset on the axial bore of the socket at the invert, showed a similar characteristic whilst Strain gauge 6, 15 mm inset on the axial bore of the spigot at the invert showed a similar but converse characteristic.

Strain gauge 1, circumferential at crown of socket, inset 10 mm showed small compressive μstrain on all tests. Strain gauge 3, 100 mm inset on the cicumferential bore of the socket at the invert registered negligible change in strain.

The local crushing pressure being applied by a 20 lb (9Kg) load through the 50 mm (2 inch) wide hanging strap is similar to the pressure resulting from a typical 39 inch (1 metre) depth of cover. The resultant circumferential bending stress for the 80 DN pipe as defined in BS EN1295-1 with this overburden is 0.16T/in^2 (2.4 MPa) with an average clay soil modulus of 0.32T/in^2 (5 MPa). This is a low level of stress because the ring stiffness of small diameter pipes is relatively high compared with large diameter pipes. Since the socket in this case is very stiff relative to the spigot, (some 25 times due to the greater wall thickness of the socket face) then if an artificial soil stiffness value of 12.9T/in^2 (200 MPa) is applied to represent side support, the stress reduces to 0.1T/in^2 (1.7 MPa) for the same load. For a less stiff 1000 DN pipe, the comparative stresses with 5 MPa and 200 MPa soil stiffness and same depth of cover would be 1.8T/in^2 (28 MPa) and 0.08T/in^2 (1.3Mpa) respectively. In other words, the effect of seal side support is far greater on the larger, more flexible pipes. It is also worth noting that according to Stantons, large diameter ductile iron pipes such as 1000DN tend to have an in-built cast ovality and the major axis is marked on the pipe. It should then be installed so that the major axis is in the crown/invert plane. Hence the overburden will tend to re-round the pipe at the horizontal axis and improve its sealing integrity. Although these tests were aimed at assessing stresses induced in the socket and spigot and not the affects on sealing capability, the tendency to re-round as a result of angular displacement from increased load, thus maintaining good seal contact around the circumference would suggest that angular deflection would not necessarily lead to joint degradation or loss of sealing performance.

The results show that the crushing stresses induced in the spigot through local seal support are much higher in this form of loading compared with a supported pipe, even with a relatively weak soil. However, they are unlikely to result in spigot failure.

The axial stress induced on a freely supported continuous section of similar pipe in pure bending under the same local loading regime would be 0.39T/in^2 (6 MPa) at the joint position. This is the same as that registered at gauge 8 with seal and maximum insertion but lower than other axial stresses recorded. However, if the equivalent overburden load was applied over the 2 metre length with 900 mm cover, which would be the case in practice, the stress would increase significantly to 4.7T/in^2. (73 MPa). In other words, the crushing and axial bending stresses induced in the spigot at the joint are significantly less than those that would be induced in the pipe barrel under pure bending.

Overall, quite high circumferential tensile stresses are induced initially in the spigot which reduce as the joint becomes seal bound. As the loads are re-distributed, the strain at the crown/invert becomes compressive. Without the seal, the resultant compressive stresses due to local high spots are of a similar magnitude with maximum insertion, and do not apparently put the pipe at greater risk. The crushing, circumferential bending stresses are significantly higher than the equivalent circumferential bending stresses for a continuous pipe. Pure bending stresses are significantly higher suggesting that small diameter pipes are more likely to fail in pure bending along the pipe barrel rather than bending or crushing failure at the joint.

JOINT FAILURE ANALYSIS AND TESTING

Joint Failure Data Analysis

A short study was undertaken in conjunction with Northumbrian Water Limited to try and understand the levels of joint failures occurring within the company, and what the major types and causes of the failures were. In the first part of the study, a brief analysis was undertaken of the general data recorded of work done on the network. The majority of the data was from sources within the distribution departments, and did not include data on trunk mains running between treatment works, pumping stations and reservoirs. The data was collated from various distribution departments including leakage find, logistics and works planning. Records were taken from December 2001 of the numbers of failures during this month, which were then broken down into pipe and joint failures where possible. Over the same period, an assessment was made of the repair fittings used by the company, to try and understand whether joint failures as opposed to pipe failures, could be identified from this source.

From the brief study of the historical data, it was found that most of the failures attended by the work crews in the three operating areas of the company were on small diameter cast iron pipe. This was not really surprising as this accounts for the vast majority of the distribution network. There were however, relatively few actual joint failures recorded. The data input system used and the works request number allocated could not always give an accurate indication of whether the incident attended by the work crew was a pipe or a joint failure. The database really only had a record of the incident, the pipe material, diameter and the location. However, on any incidents attended involving larger diameter cast iron pipe e.g. above 12 in., the majority of the failures, where identified, were due to leaking joints. From the total number of failures recorded during this period and analysed, it was estimated that the number of actual joint failures was around 8% of the total.

In the second part of the study, a field assessment was undertaken in one distribution area of the company over a three-month period. The area has a good mixture of urban and rural locations and should have provided a fairly balanced view of the distribution network. The largest number of mains are located in this area, and traditionally it has higher numbers of bursts than the two other areas. The general number of all pipeline failures was documented, and using a failure data pro-forma, the characteristics and reasons for particular failures was also recorded, when the site could be attended by supervisory staff. Where possible, if a joint failure was identified, a site visit was made to record any useful supplementary information and if possible to retrieve a pipe sample. The data collected from these site visits over the particular time period is summarised in Table 4.11.

Table 4.11
Summary of results from NWL field failure data

Material	Pipe dia	Failure type (mains)	Failure type (fitting)	Excavation reason	Comments
CI	2 in.	Circumferential crack		Leak survey	
CI/lead	3 in.		Ferrule	Leak survey	Ferrule failed on connection to service
CI	4 in.		Ferrule		Hole formed ferrule screw to main
DCI	10 in.		Leak past flange gasket	Planned	Flange fitting retained for examination
CI/PVC-U	3 in.		PE stub flange	Failure	PVC-U cut short resulting in joint failure
CI	18 in.		Push-fit joint	Leak survey	Photographed
PVC-U	8 in.		Joint	Failure	Split along length of joint
Lead	25mm	CI corrosion		Leak survey	
Lead	25mm	CI corrosion		Leak survey	Lead service severely degraded
CI	3 in.	Circumferential crack		Leak survey	
CI	4 in.	Circumferential crack		Leak survey	
CI	4 in.	Circumferential crack		Leak survey	
CI	4 in.	Circumferential crack		Failure	
Steel	4 in.	Circumferential crack		Leak survey	
CI	3 in.	Longitudinal split		Leak survey	
PVC-U	25mm	Joint to PE pipe		Failure	
PVC-U	4 in.	Socket and spigot leak		Failure	Joint retained for analysis
CI	6 in.		Leak past flanged gasket	Planned	Taper from 6 to 4". Bolt head snapped
CI	3 in.	Circumferential crack		Leak survey	
CI	4 in.	Circumferential crack		Leak survey	Pipe appeared to be in good condition
CI	3 in.		Ferrule	Leak survey	Ferrule detached from CI main
AC	4 in.		Ferrule	Failure	
PE80	25mm		EF leak	Failure	
PE80	25mm		Joint to boundary box	Leak survey	

(continued)

Table 4.11 (continued)

Material	Pipe dia	Failure type (mains)	Failure type (fitting)	Excavation reason	Comments
PE100	90mm		VJ coupling leak	Leak survey	Main relined 18 months ago. Joint not fitted correctly
CI	6 in.	Circumferential crack		Failure	
PE100	90mm	PE pipe wall		Leak survey	PE bend between two CI lengths
PE100	90mm	3rd party damage		Leak survey	New section fitted
CI	6 in.		Ferrule	Failure	
CI	4 in.		Leak past stub flange gasket	Failure	Cut out for analysis
PE80	90mm		Valve/hydrant body or gland	Leak survey	Bolts replaced
Lead	25mm	Split		Failure	
CI	4 in.	Longitudinal split		Failure	Main known to burst frequently
CI	12 in.		Leak past flange gaskets	Leak survey	
AC	4 in.	Longitudinal split		Planned	New section fitted
CI	6 in.	Longitudinal split		Leak survey	
Lead	25mm	Split		Leak survey	Service under road
CI	18/24 in.		Leak past flange gaskets	Leak survey	

Courtesy Northumbrian Water Ltd.2003

The most common failure type recorded was for circumferential cracks on small diameter cast iron pipes. This is due to the composition of the NWL distribution network with these size pipes being the most frequently occurring diameter. Circumferential cracks followed by longitudinal splits account for most of the pipe barrel failures. The most common type of joint failure was at the connection of the main with the service pipe. Corrosion of the flanged gasket joints and displacement of the rubber seal was the most frequent reason for failure on the remaining joints.

The pipe joint failures tended to occur on the larger diameter mains. This was indicated during the course of the study, and was the general view held by the network operatives. Degradation leading to displacement of rubber gaskets and the corrosion of bolts were the most common causes of joint failures. Due to the expense and impracticalities of removing, rather than repairing these large diameter joints, laboratory examination of joints was not possible. However, some of the smaller diameter joints were removed for further examination, as described earlier.

DESTRUCTIVE JOINT TESTING

There is little reference material available on the causes of cast iron joint failure in water pipes. A number of references cover testing of lead yarn spigot and socket joints, primarily from gas mains. These include references from Attewell and Singhal and consider the loads required to deflect or pull joints, but do not generally describe the manner in which the joint fails. The objective of performing the destructive tests described below was to observe the way in which joints fail under different loading regimes and to determine whether such tests could be used to indicate parameters that could be measured to allow prediction of future joint failure. Three joints were tested, each under a different loading regime. The joints were 6" Cast Iron lead yarn spigot and socket. Two were subjected to tensile tests and the third to a bending test. The tests were performed by Bodycote Materials Engineering Limited (Daventry, UK).

An issquared Ltd representative was present at the tests. The test procedure originally issued to Bodycote was modified slightly as the tests progressed, to accommodate practical considerations. A test report was produced by Bodycote and this is included in Appendix J.

Tensile Test 1 (Stepped)

The stepped tensile test consisted of pulling the joint apart in displacement increments, allowing the joint to rest for a period of time between displacements. The test continued until significant leakage occurred.

The pipe section including the joint was cut to approximately 17.3 in. (440mm) in length with the joint at the centre. The ends of the section were sealed to hold 14 psi (1bar) internal water pressure and prepared for connection to the test machine. See Bodycote Test Report in Appendix J for details of test set up and test procedure.

Figure 4.26 shows the joint assembled into the testing machine in preparation for the stepped tensile tests. The lower part of the test sample is hidden from view by the white plastic container, which was used to catch leaked water. The end of the socket is just visible above the white container. The pipe work connected to the spigot is the water feed to the joint and the pressure transducer is connected to the left of the pipe work.

The joint was clamped in position before water pressure was applied, to eliminate axial displacement due to the water pressure. The axial force imparted by the water pressure was approximately 2kN.

Figure 4.26 Joint assembled in testing machine for stepped tensile test

Stepped Tensile Test Observations

Figure 4.27 shows the first displacement steps. A force of approximately 14kN (+2kN from internal pressure) was required to move the joint initially. See Bodycote Test Report in Appendix J for complete data plots.

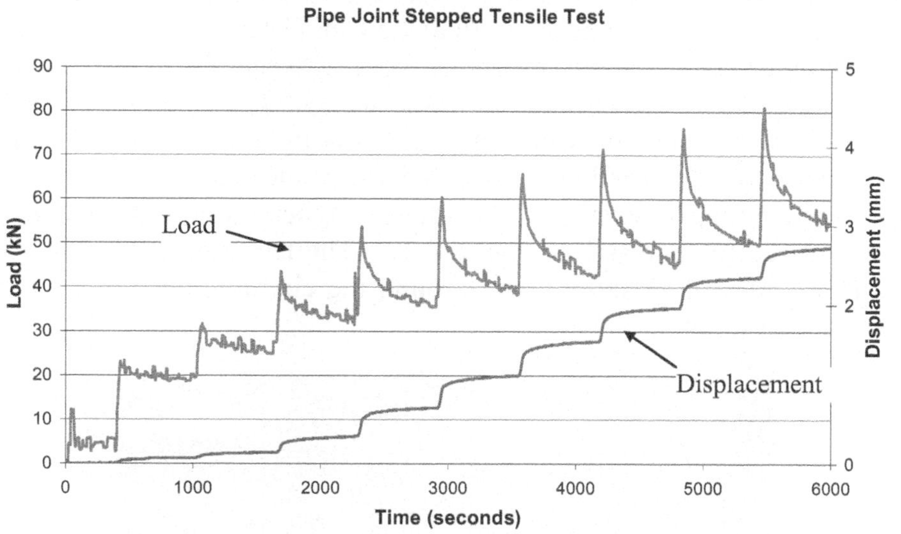

Figure 4.27 Plot showing first few displacements in the stepped tensile test

When water pressure was applied at the start of the test, two very small weeps were observed at the joint, Figure 4.28. However, one of these sealed itself after 0.06 in. (1.5mm) extension and the other sealed itself at 0.157 in. (4mm) extension.

Figure 4.28 Slight weep at start of stepped tensile test

It is assumed that the yarn had dried slightly since the section was removed from the ground and this may have caused the slight leak. The joint was water soaked for two days prior to the test in accordance with the issued test procedure but this soak period may not have been long enough.

At 0.55 in. (14mm) extension, a slight seepage of water between spigot and lead was observed and this remained fairly constant until at 1.54 in. (39mm) small trickles were observed. It was observed that the spigot was pulling out of the lead except for in one place where the lead had adhered to the spigot and was pulling out of the socket. The small trickles observed at 1.54 in. (39mm) were not at the position where the lead was pulling out of the socket. At 1.65 in. (42mm) water seepage was observed at the position where the lead was pulling out of the socket and this increased to a strong flow at 1.77 in. (45mm). When the lead pulled clear of the socket the joint was considered to have failed, Figure 4.29. The test was stopped at 1.85 in. (47mm) extension. Figures 4.30 to 4.33 show views around the circumference of the failed joint after removal from the testing machine.

Figure 4.29 Joint failure at 1.85 in. (47mm) extension

Lead adhering to spigot

Figure 4.30 View of the failed joint after removal from the test machine (0 deg)

Figure 4.31 View of the failed joint after removal from the test machine (90 deg)

Figure 4.32 View of the failed joint after removal from the test machine (180 deg)

Figure 4.33 View of the failed joint after removal from the test machine (270 deg)

Figure 4.34 shows a section through the failed joint, through the plane in which the lead pulled out of the socket. A steel backing ring 0.59 in. (15mm) wide by 0.20 in. (5mm) thick can be seen near the deepest part of the socket. The inner surface of ring was in good condition, with evidence of bitumen on it. It is believed that this ring would have been push fitted onto the spigot end prior to assembly of the joint gap, to hold the yarn in place. The gap between the ring and the socket was completely full of corrosion products, not from the ring, but similar to that adhering to the inner wall of the pipe. The backing ring had been pulled towards the end of the socket by about 0.39 in. (10mm) on the major lead extrusion side and 0.08 in. (2mm) on the opposite side.

Figure 4.34 Section through joint from stepped tensile test

The maximum lead pull out from the socket was 1.42 in. (36mm), on the left side of Figure 4.34. At this position, the length of lead was 1.65 in. (42mm) and the yarn 0.59 in. (15mm) up to the backing ring. On the opposite side of the joint, right of Figure 4.34, minor lead pull out of 0.16 in. (4mm) was observed. At this position the length of lead was 2.17 in. (55mm) and the yarn 0.59 in. (15mm) up to the backing ring.

As the contact area between the spigot and lead is significantly less than that between the lead and the socket, it would be expected that the spigot would pull out of the lead, leaving the lead in the socket. This is particularly likely to occur in joints where the socket contains a recess for the lead, as in this example. In this case, it is suspected that the backing ring and compressed yarn had started to push the lead out of the socket before the backing ring had come free of the spigot, thus breaking any bond between lead and socket, and causing the lead to plastically deform in the region of the recess in the socket.

The spigot was pulled away from the lead after sectioning the joint, and a layer of bitumen was present at the lead/iron interface. This is shown in Figure 4.35.

Figure 4.35 Section through joint showing bitumen at lead/iron interface

It is considered that this bitumen layer would have been present on the spigot at the time of jointing, and would melt when in contact with the molten lead, but would be unlikely to burn away due to the absence of oxygen. This bitumen layer may well improve the lead/iron adhesion in this type of joint.

Tensile Test 2 (Cyclic)

The cyclic tensile test consisted of displacing the joint axially, allowing it to rest for a period of time, pushing it back to its original position and again allowing it to rest for a period of time. This was repeated for a number of cycles, the displacement increased, and then the process repeated. The joint was subjected to 50 cycles at displacements of 0.2mm and 0.4mm, and 20 cycles at 0.6mm, 1.0mm and 2.0mm.

The sample was prepared in the same way as for the stepped tensile test and assembled into the same test machine. See Bodycote Test Report in Appendix J for data plots and details of test set up.

Cyclic Tensile Test Observations

Figure 4.36 shows load versus displacement for the first six cycles of the test.
Joint displacement was measured at two places on opposite sides of the joint and the average displacement is shown in Figure 4.36. The test was paused after 13, 0.016 in. (0.4mm) cycles, due to time constraints, and then restarted. At the restart, the difference between the two displacement measurements was minimal and remained so for the remainder of the test. The initial difference between the two displacement measurements could indicate that the joint was

slightly misaligned, or that there was less resistance to movement on one side of the joint than the other.

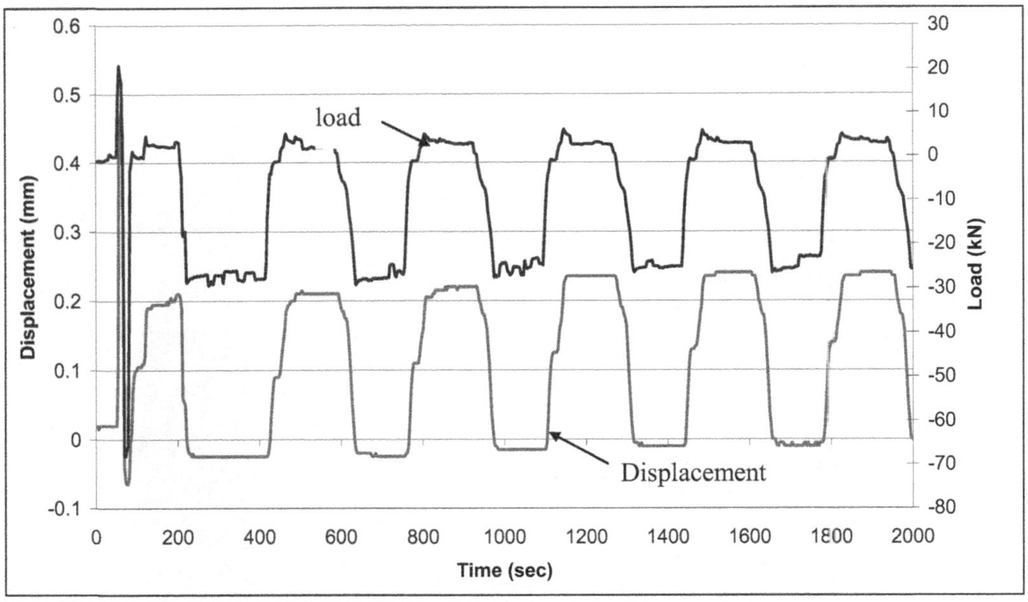

Figure 4.36 Plot showing first few cycles of 0.008 in. (0.2mm) Cyclic Tensile Test

As with the stepped tensile test, very small weeps were observed at the joint when water pressure was applied at the start of the test. After five, 0.008 in. (0.2mm) cycles, the weeps were only observed on the compression half of the cycle and sealed on extension. After 19 cycles the leak was slightly worse but still only on the compression half of the cycle (see Figure 4.37).

Figure 4.37 Joint after 19, 0.2mm cycles and after 2 minutes rest on compression

The leak rate was only slightly higher during the 0.016 in. (0.4mm) displacement test but increased to continuous dripping during the 0.024 in. (0.6mm) test. After 10 cycles of 0.04 in. (1mm) displacement the dripping increased to running (see Figure 4.38). No increase was observed for the remainder of the 1mm and the whole of the 0.08 in. (2mm) test. Note that in Figure 4.38, aluminium tape has been applied to the outside of the socket to channel the leak water for observation. In general, the leak rate increased slightly during the compression part of the cycle. The test was terminated after 20 cycles of 0.08 in. (2mm) displacement.

Figure 4.38 Leak water running continuously

Figures 4.39 and 4.40 show sections through the joint. A 0.275 in. (7mm) thick cement mortar lining was present, and it can be seen that this had fractured in line with the end of the spigot. Although the fracture was not observed before sectioning the joint, it was difficult to see inside the spool containing the joint before sectioning, and therefore it is possible that the fracture had occurred before the test.

There were different amounts of lead and yarn on each side of the joint; 2.28 in. (58mm) lead and 1.26 in. (32mm) yarn on one side; 1.89 in. (48mm) lead and 1.65 in. (42mm) yarn on the other. This difference could result in a variable resistance to axial movement between one side of the joint and another, and hence explain the observed difference in displacement measurement on opposite sides of the joint. There was no evidence of lead extrusion or pull out at any point around the circumference.

The lead to spigot interface appeared to have broken on one side of the joint, potentially causing a leak path. This position was consistent with the location of the leakage shown in Figure 4.37.

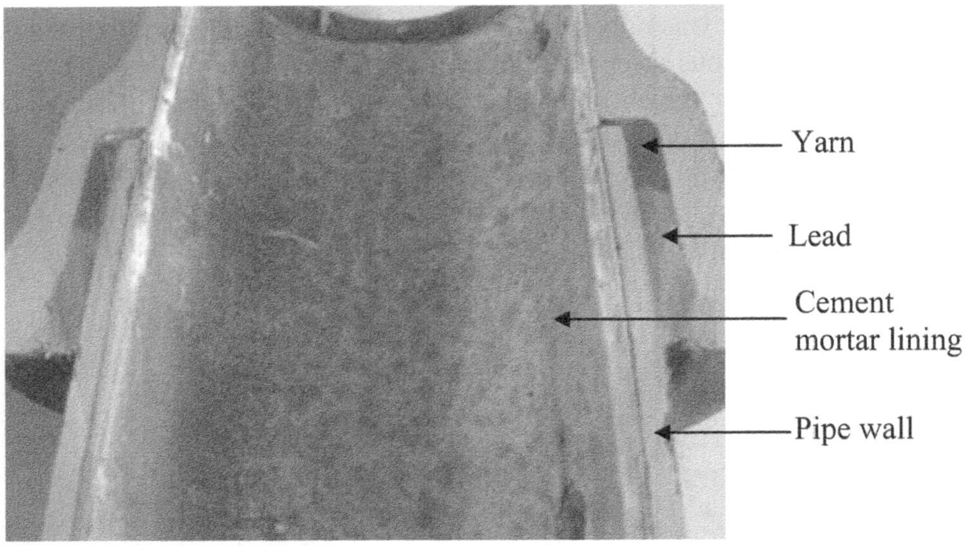

Figure 4.39 Section through joint from cyclic tensile test

Figure 4.40 Section through joint from cyclic tensile test

Bending Test

The bending test was carried out in two stages, since it became necessary to move from one test machine to another partway through the test. This was because the force required to bend the sample became higher than the 90kN capacity of the first machine. See Bodycote Test Report in Appendix J for details of test set up.

The pipe section was set up for the first part of the test as shown in Figure 4.41. The distance between the pipe supports was 47.24 in. (1200mm) and the distance between the load points was 11.81 in. (300mm). Pressure was maintained manually and was not recorded. Axial displacement of the joint due to the force exerted by the internal water pressure was prevented by the studding along the length of the sample, which was connected to blocks at each end.

A displacement measurement pad was bonded to the socket (see Figure 4.42) and this was contacted by a linear variable displacement transducer (LVDT), which was supported on a magnetic base.

Figure 4.41 Bending test set-up

Figure 4.42 LVDT measurement point

For the first part of the test, the joint was displaced by 0.1 degree increments (0.02 in. linear displacement) at a displacement rate of 0.04 in. (1mm) /minute, every 10 minutes. For the second part of the test, the joint was displaced in 0.04 in. (1mm) increments. Load and displacement were recorded against time.

Bending Test Observations

A slight weep was observed at the start of the test (Figure 4.43, 1 drip in 45 seconds) and this increased immediately when the joint was displaced.

Figure 4.43 Slight leak at start of test

Figure 4.44 shows drip rate versus angular offset for the first part of the test. As far as could be observed, the leaks were from the sides of the joint (pivot points) rather than from the crown or invert.

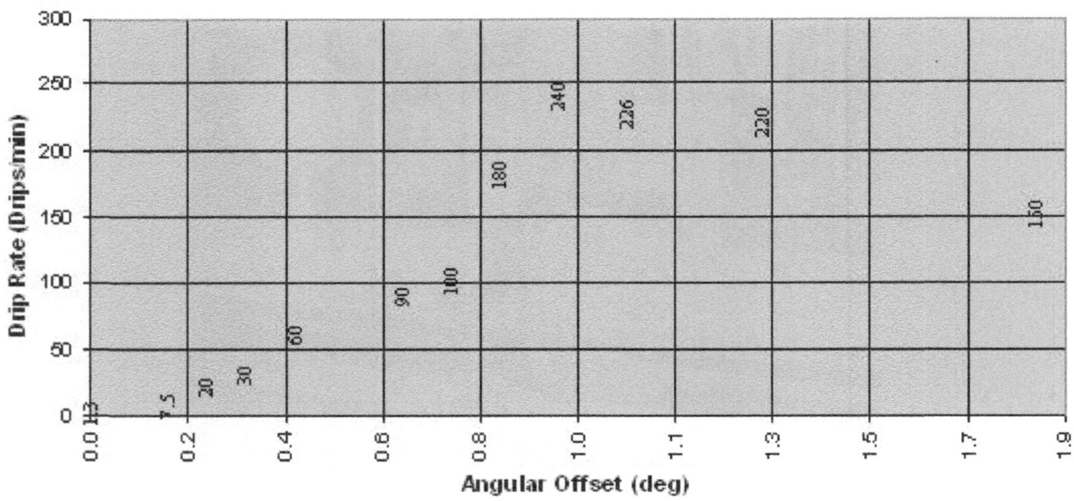

Figure 4.44 Drip rate for the first part of the bending test

At the end of the first part of the test after 0.38 in. (9.7mm) displacement, 1.9 deg angular offset, the load was removed and the joint allowed to relax. When fully relaxed the measured displacement was 0.31 in. (8mm) (1.5 degrees) and this value was used as the start point for the second part of the test.

The sample was moved to a higher capacity testing machine and set up in a similar manner as in the first part of the test. The only difference was that because larger loads were expected, soft load-spreading pads were applied under the load points (see Figure 4.45) to reduce the possibility of local fracture at these points.

Figure 4.45 Bending test set up for second part of test (higher capacity testing machine)

The test was continued in the same way but with displacement steps of 0.04 in. (1mm or 0.2 degrees). The leak rate increased as shown in Figure 4.46, but between 10mm and 15mm displacement (1.9 degrees and 2.9 degrees) the drip became a continuous trickle. The drip rate reduced slightly towards the end of the test.

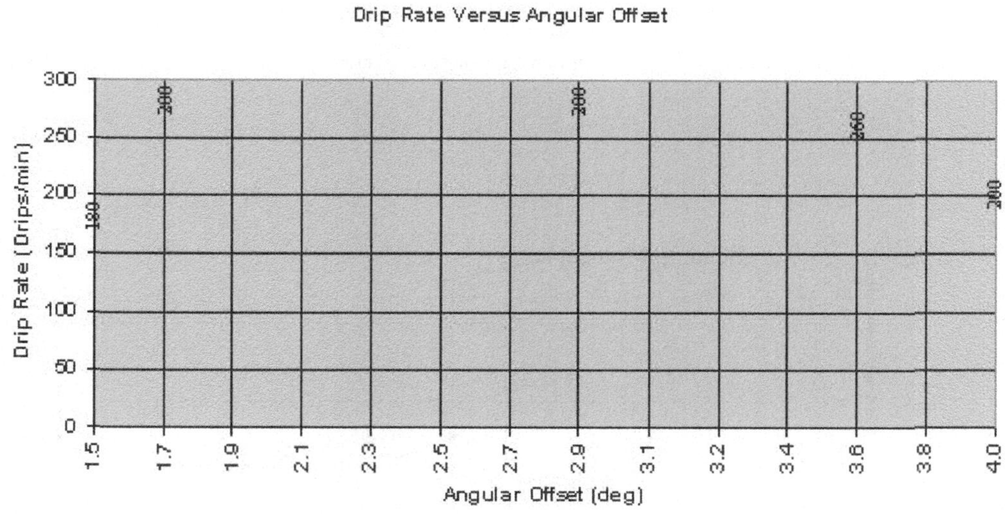

Figure 4.46 Drip rate for the second part of the bending test

At 0.83 in. (21mm) displacement (4 degrees), the pipe fractured at entry to the socket (see Figure 4.47). This occurred during a hold period, approximately 7 minutes after a displacement.

At this time, the bending moment had reduced from a peak of 33kNm at the end of load application to 30kNm.

Figure 4.47 Failed pipe showing spigot fracture at entry to socket

The Bodycote Test Report in Appendix J shows data plots in the form of applied load (kN) and displacement (mm) versus time. Figures 4.48 and 4.49 show bending moment (kNm) and angular displacement (degrees) versus time. It can be seen from the figures, that the load required to deflect the joint remains reasonably consistent for each deflection increment over the duration of the test. There is a clear reduction in load during the hold period between displacements, due to relaxation of the lead. Figure 4.48 shows that the extent of relaxation after each displacement increases significantly during the first part of the test. This pattern is repeated for the second part of the test, as shown in Figure 4.49.

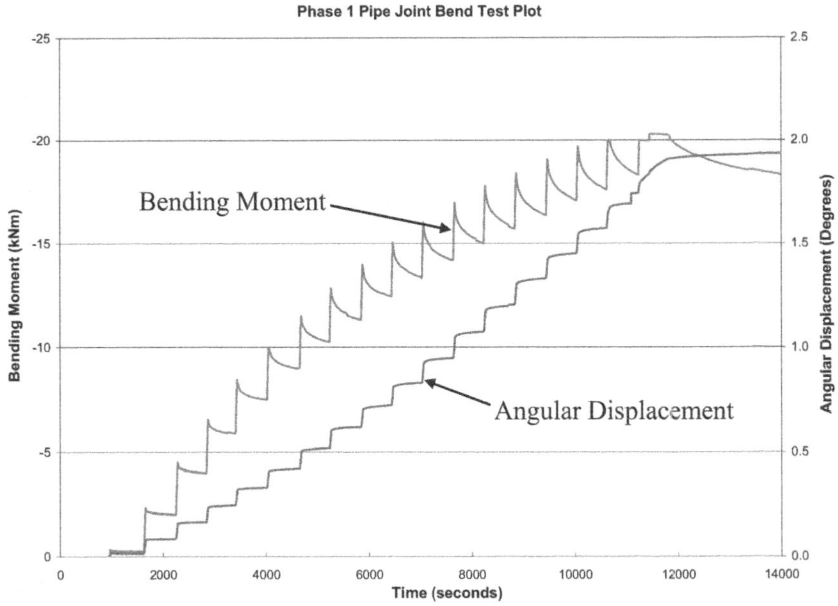

Figure 4.48 First part of bending test

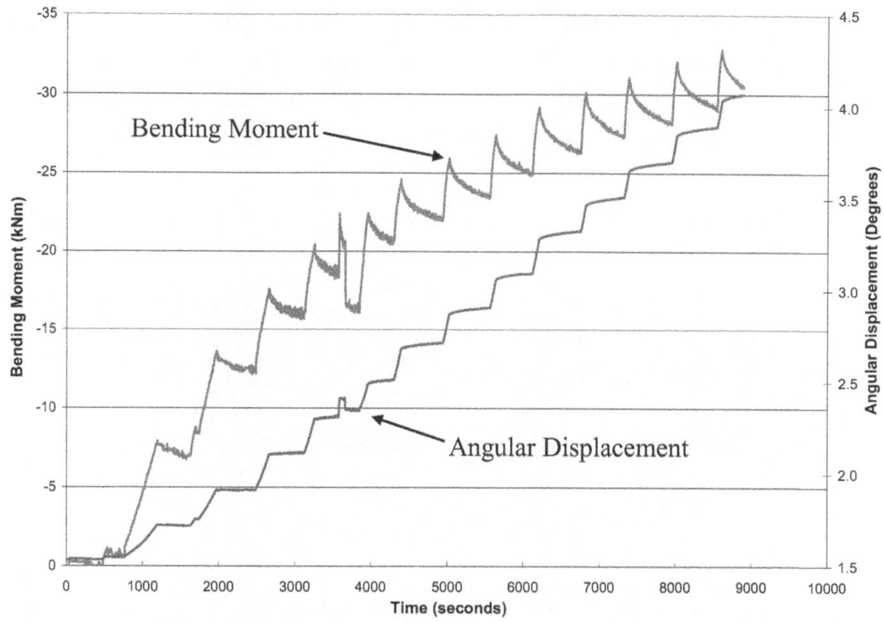

Figure 4.49 Second part of bending test

Figure 4.50 shows a section through the joint, in the plane of bending. The right hand side of the figure shows the invert of the joint during the bending test. On this side, the end of the spigot has moved towards the outside of the socket and pulled out of the socket by about 0.24 in. (6mm). A 0.12 in. (3mm) gap can be seen where the spigot has pulled away from the lead. On the left hand side (crown during test), the end of the spigot has moved away from the outside of the socket and pushed into the socket, with probable metal to metal contact between the end of the spigot and bottom of the socket. The lead on this side has been compressed.

There were different amounts of lead and yarn on each side. 1.97 in. (50mm) of lead and 1.5 in. (38mm) of yarn on the left side; 1.77 in. (45mm) of lead and 1.65 in. (42 mm) of yarn on the right side (Figure 4.48).

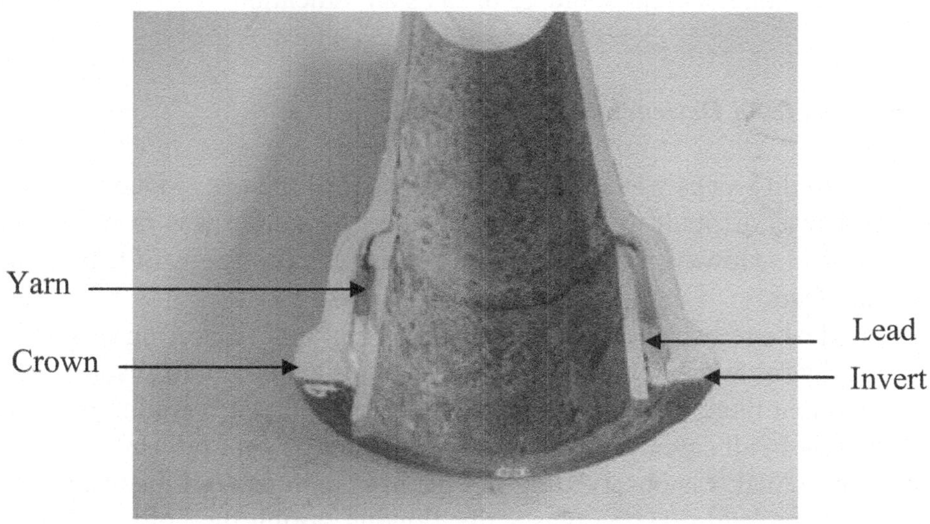

Figure 4.50 Section through joint from bending test (in plane of bending)

Figure 4.51 shows a further section through the joint, at 90 degrees to the plane of bending. This was the point of rotation of the joint during the bending test, and the point at which most leakage was observed. However, visual inspection of the sectioned joint did not provide any indication of an obvious leak path.

Figure 4.51 Section through joint at 90 degrees to plane of bending

Destructive Joint Testing Conclusions

This section provides conclusions on the destructive testing of lead yarn joints, together with suggestions of parameters that could be used as an indication of the onset of failure for each loading regime.

General Conclusions from Destructive Joint Testing

In all tests, slight weeps were observed at the start of the tests. Due to wetting of the area around the weep, it was difficult to determine whether the source was from between spigot and lead or socket and lead. However, it was considered that the first signs of leakage were generally between spigot and lead.

In all tests, leakage became significant during the test. Referring to the classes of joint degradation shown in Table 4.2, leakage was Class 5 to 6 (significant to severe).

It was noted that there were different depths of lead and yarn around the circumference of all of the joints tested. In all cases, leakage was an indicator of joint failure.

A reduction in load was observed during the hold periods after displacements for all tests. This was a result of the lead relaxing/plastically flowing during the hold periods. From all tests undertaken, it appears that the lead maintains its ability for plastic flow, and to some extent seal heal, even at a considerable age.

Further testing of a larger number of joints, in a range of diameters would provide a better understanding of the way in which the joints fail under different loading regimes, and hence, an indication of the physical parameters that can be measured to provide an indication of the onset of failure. An appropriate test programme should be devised and undertaken.

Conclusions from Tensile Test (Stepped)

The spigot pulled out of the lead, except over a segment extending around 30% of the circumference of the joint where the lead pulled out of the socket. This was most likely due to the presence of a backing ring adhered to the end of the spigot.

From this evidence, it is concluded that the lead will pull out of the socket where cast iron pipes have been manufactured with an external raised step at the end of the spigot, and that the spigot will pull out of the lead where pipes do not have the external raised step.

A small weep was observed at the start of the test. Leakage became significant after 1.54 in. (39mm) extension, in a location away from where the lead had pulled out of the socket, and at the position where the lead had pulled out of the socket after 1.65 in. (42mm) extension. The test was terminated after the lead had pulled clear of the socket.

Potential measurement parameters indicating the onset of failure due to axial pull out: spigot withdrawal, lead withdrawal, disbonding of interface between lead and iron, leakage.

Conclusions from Tensile Test (Cyclic)

A small weep was observed at the start of the test. Leakage became significant during the 0.6mm cycles. Leakage was generally worse on the compression part of the cycle.

Potential measurement parameters indicating the onset of failure due to cyclic axial displacement: disbonding of interface between lead and iron, leakage.

Conclusions from Bending Test

A small weep was observed at the start of the test. Leakage became significant after 0.2 degrees angular displacement. The test was terminated when the joint fractured circumferentially at the point where the spigot entered the socket.

The observed leakage path was at 90 degrees to the plane of bending i.e. at the expected points of rotation of the spigot in the socket. The location of weeps on an exposed joint may therefore give an indication of the plane of bending, and thus be useful when assessing cause of failure.

Potential measurement parameters indicating the onset of failure due to bending: Variation in joint gap around circumference, angular displacement, spigot and socket becoming metal bound, disbonding of interface between lead and iron, leakage.

FINITE ELEMENT ANALYSIS (FEA) MODELLING

Introduction

As part of the condition assessment process, stress analysis of strategic pipe crossings and their related structures is undertaken. It was therefore considered that this approach could be used for the testing and possible condition assessment of joints in water distribution pipeline systems in a similar manner. This section reports on the evaluation of the modelling approach for assessing pipe. Concrete joints were selected since there is a relatively lengthy network of large diameter concrete pipes and a number of joint rehabilitation programmes are known to be in place. Lead yarn socket and spigot joints were selected since they represent the largest population of joints of all sizes. A non-linear module was used in order to handle the complex material and interaction phenomena encountered in these joint configurations.

The evaluation of FEA software was principally conducted by building a model of a socket and spigot joint and subjecting this to sensitivity studies in order to estimate the significance of material properties, geometry and loading variation. Some 33 different models were constructed, and 639 trial analyses carried out.

The reporting of this work is detailed in Appendix K and comprises:

- Introduction to the finite element method
- The objectives of the research
- The method of building the model (covering geometry, material properties including creep, loading and other issues such as contact)
- Results of the sensitivity studies

From this work, the following conclusions and recommendations are made for further modelling investigations, tests which could be carried out, and advice on the use of the technology.

Conclusions

- The initial configuration of a joint (alignment, lead geometry etc) is key to its subsequent condition.
- Material properties are not generally the dominant factors in joint behaviour, although interfacial characteristics such as adhesion are more important.
- There may be significant circumferential variation in joint behaviour, depending on the loading regime.

On the basis of these, several points may be made on the success of this modelling technology.

- When a problem of interest arises, it can significantly aid understanding if detailed measurements of pipe joint geometry are made.
- Finite element analysis has the potential to add a great deal to the understanding of joint behaviour, when suitable input data are available, adequate computing power and a model of sufficient complexity for handling relevant phenomena such as

multiaxial contact (misalignment), and can give good agreement with observations of circumferential variations, leak behaviour, importance of good installation etc

On the basis of these points, several recommendations can be made for further investigation and the use of this technology.

Recommendations

The findings suggest that the most important feature to be quantified on site is the lead and joint geometry, and that the most important aspect of joint condition meriting laboratory testing is the interfacial properties – how the whole joint sticks together. Subsequent developments should include quantification of the loads on any joint of interest, e.g. by taking levels near any pipeline where ground movement and/or leakage is suspected, and strain-gauging critical pipe sections.

It is important that research be carried out in the correct sequence. For example, there would be little point in demonstrating that stresses in steel can be very sensitive to changes in the elastic modulus since in practice this hardly varies at all. A key item to test for in lead is not, for example, the absolute value of the stiffness or hardness *per se* but rather the *range* of values and the variability, i.e. the frequency distribution. It would therefore be useful to carry out a programme of integrated imaging, pull-apart and subsequent material testing on sets of small diameter cast iron pipe joints which have been removed during intrusive assessment or replacement in order to identify the range and variability of properties.

Some of the above results may only become truly operationally meaningful when the variability of the inputs in practice is known.

Further Research

- Apply the technique to other joints e.g. concrete, mechanical couplings
- Use other, more complex codes with more sophisticated contact models e.g. ABAQUS
- Refine models in areas of interest (e.g. meshes around contact interfaces)

Tests

- Friction/adhesion properties
- Strain-gauge joints under push/pull/twist

Use of Technology

- Seek expert modelling advice with complex modelling technology (eg sophisticated contact capability) for specific problems
- Build up confidence in modelling technology through benchmarking and uniaxial linear behaviour of simple structures
- Build up confidence in modelling technology with laboratory tests of input properties and verification of overall structural performance as above

CHAPTER 5
POTENTIAL TECHNIQUES FOR THE DETECTION OF JOINTS

INTRODUCTION

This section describes techniques identified from the initial technology survey as having the potential to detect joints. The survey covered patent, internet searches and company contacts. Novel, developing, emerging and representative current techniques are considered and described. Benefits, limitations and conclusion for each technique are provided whether they are considered to have potential for this application or not. Field trials and laboratory tests were carried out on selected techniques, that showed potential but for which there was little information available on their performance in this application. These activities are described more fully in this section. The same format is applied in Chapters 6 and 7 on condition assessment.

NON INTRUSIVE - SURFACE

Magnetic Detection

All steel and cast iron pipes have a locked-in 'passive' magnetic field created at the time of manufacture, illustrated in Figure 5.1.

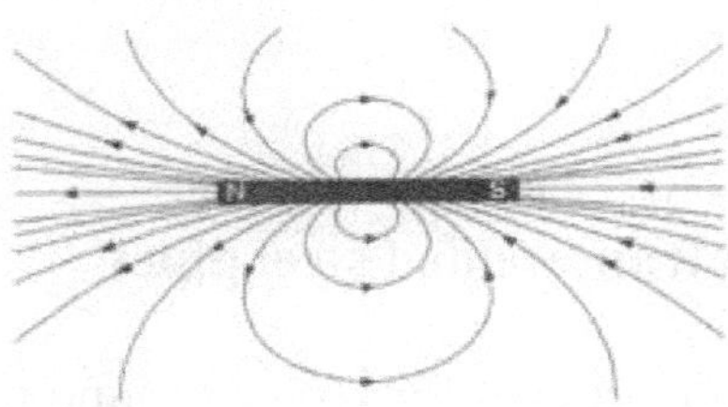

Figure 5.1 Single pipe magnetic field

This feature has led to the development of equipment capable of tracking pipelines. In the case of steel pipelines, the magnetic field is locally destroyed by the welding process and hence there is discontinuity at the weld. Vehicular machines have been developed for both sub-marine and above ground applications to track a pipeline and in the case of welded steel, identify the joint due to the loss of the pipes magnetic field, Figures 5.2 and 5.3.

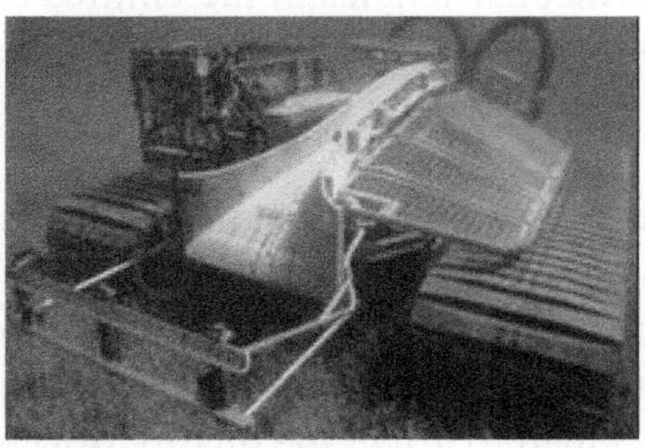

Courtesy Innovatum
Figure 5.2 Sub-marine pipeline tracker

In practice, the equipment picks up both the pipe and the earth's magnetic field and a resultant field is produced. Typically the resultant field is not directly above the pipe but adjacent to it. In order to compensate for the earth's magnetic field, two sets of sensor arrays are used and by considering the difference in distance, the influence of the earth's magnetic field is identified.

This principle will apply equally to cast iron socket and spigot installations, each spool has its own field as illustrated in Figure 5.1, such that there is a null field at the joint due to the discontinuity at the joint. A similar effect for steel is shown in Figure 5.3 as the pipeline tracker moved along the pipeline. Hence, joints can be identified by noting this position where the field from one pipe diminishes, before the signal is received from the next.

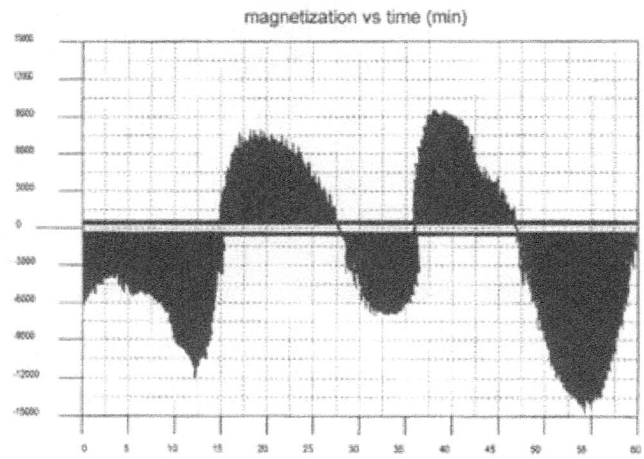

Courtesy Innovatum

Figure 5.3 Actual magnetic field from welded steel pipes

A hand held tool has been developed for above ground use by Radio Detection Equipment (UK) called the Magnatrak. The tool comprises of two coils set apart in the body of the tool, which emit a noise signal to assist in identifying peak output and a digital display relative to the strength of the magnetic field. The tool incorporates a sensitivity and volume control. The company advised that the tool requires a certain amount of experience in use of the sensitivity control and positioning of the tool. Care has to be taken regarding nearby metallic items, such as steel toe-caps, buckles, coins, pipes, vehicles and street furniture such as manhole covers.

Initial tests were not encouraging due to inexperience and difficulty in interpretation of the signals. There appeared to be inconsistencies in the emitted signal however, once the effect of the earth's magnetic field had been understood, then the tool was found to be more effective. In many cases, it was necessary to track the pipe off centre and by careful control of the sensitivity, joints could be identified.

The tool can be held vertically or horizontally or at any angle, and the best angle for different pipes and site conditions, was found by trial and error. The signal output is influenced by depth of cover and pipe diameter. At over 6 ft 6 inches (2 metres) of cover on a ductile iron 48-inch (1200 mm) diameter pipe, it was difficult to detect the joints with confidence. Also, the direction of following the pipe seemed to affect the quality of signal.

Overall, as the operator gained experience, confidence in the capability of the tool increased. However, its major limitation is that it cannot be used effectively in the vicinity of other metallic pipes, street furniture (e.g. manhole covers) vehicles etc., and is not therefore suitable for congested urban environments.

Nevertheless, the Magnatrak tool is effective as a pipe locator/tracker and in favourable conditions will detect joints. The device clearly has some potential and is worthy of consideration where joints need to be located from above ground. The principle of utilizing a pipes inherent passive magnetic field can be adopted as a viable means of tracking pipes and identifying joints. Use of the system can be summarised as follows:

- Benefits of the system are that it is light, portable and can identify joints in a rural environment with experience
- Limitations of the system are considerable interference from other metallic street furniture and vehicles, and the depth of mains
- In conclusion, there is potential for joint detection in a rural environment but current usage is unknown

Ground Penetrating Radar (GPR)

This technique based on radar, is designed to locate buried pipe under layers of different materials. Pulses of electromagnetic energy are launched into the material under investigation. These propagate through the ground and are reflected by the sub-surface target or by interfaces between different media. The radar measures the time taken for a pulse to travel to and from its target, which indicates its depth and location. The transit time is a function of the dielectric properties of the material. Some materials are easier to penetrate than others. Commercially available GPR systems allow three-dimensional mapping of underground utility infrastructure, including pipes and cables. Data from GPR surveys can be provided in CD format and is exportable to GIS. Examples of different forms of GPR equipment and a typical output are shown in Figures 5.4 and 5.5.

There are two theoretical mechanisms by which ground penetrating radar (GPR) might be capable of locating joints in metal water pipes. It is emphasised that these are theoretical approaches. They have a sound basis in theory, but have not to the author's knowledge been demonstrated to work in real situations.

Courtesy Tricorn Technologies

Figure 5.4 Prototype hand operated system

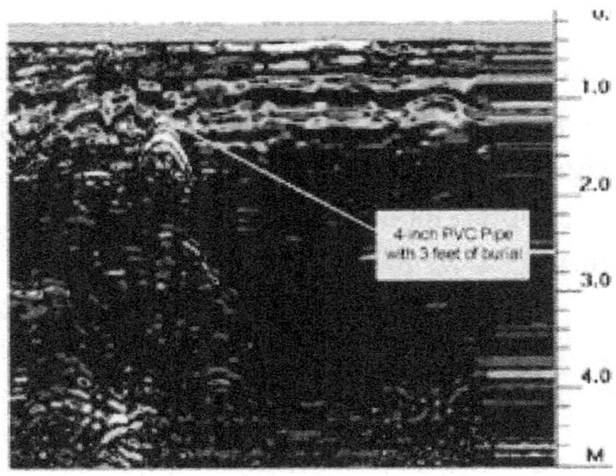

Courtesy Tricorn Technologies

Figure 5.5 Example of a 4 in. PVC pipe, buried at a depth of 3 feet

The first possibility relies upon the increased mass of metal (the "bulge") in the vicinity of a joint, e.g. a bell-and-spigot joint. This larger mass of metal will produce a larger radar echo from the pipe, referred to as an increase in the "radar cross-section" of the pipe. An increase in diameter of the order of 10-20% of the pipe diameter would cause significant increases in the echo strength from the pipe. Thus, by tracking the pipe with radar, and noting the points at which the echo strength increases, joints might potentially be identified.

The important caveat with this approach, however, is that the echo strength from the pipe will also be varying away from joints. This arises because of local variations in soil conditions (pockets of moister or dryer soil, different aggregate size, etc.), and because of fading caused by interference in the reflected waves from inhomogeneities. Thus, unless the soil were very uniform, it could be difficult to know whether a particular echo strength variation was due to a joint or to soil variations.

The second possibility would be the employment of a high-frequency (thus high resolution) GPR, of sufficiently short wavelength of operation that the "step" in the pipe caused by the joint bulge could actually be imaged. The step produced by the bell of a bell-and-spigot, of a few cm in size, would in principle constitute a corner reflector to a radar of say, 1 to 2 GHz centre frequency. This would produce a strong, distinctive return, and further processing could in principle produce an image of the joint, albeit of low resolution.

The drawback here is that radar's of such high frequency would, owing to the very high losses in moist soils at high frequencies, be very limited in their depth of penetration. In typical British soils, perhaps no more than 1 ft 8 inches (0.5 m) might be expected. In drier and more uniform soils (e.g. U.S.A. or Middle East), there would be improved potential for this approach.

It may therefore be possible to develop a GPR with the capability to detect joints in metallic water pipes. However, searches performed as part of this study have not revealed any commercially available GPR systems with sufficient resolution to allow the detection of joints. This is thought to be because the increase in effective pipe diameter in the vicinity of a pipe joint would not be detected above the expected noise in the reflected radar signals. In summary:

- The system could provide accurate location of pipe joints
- The major limitation is that the local variation in soil conditions would affect performance

- In conclusion, there are no known commercially available systems for joint detection, and no known research currently in this area.

Pipe Locator and Direct Measurement

This is a well-established technique widely used in the Gas Industry for joint encapsulation rehabilitation projects. The pipe and pipeline are identified using a standard pipe and cable locator. An initial excavation is carried out to locate the first joint. Thereafter other excavations are prepared based on the spool length of the buried pipe. Due consideration of bends and short pieces should minimise aborted excavations. Despite its limitations, it is cost effective in the absence of any superior technique. Use of the system can be summarised as follows:

- Benefits are that operationally it can find joints
- Limitations are that it requires the skill of the operator and success depends upon the judgement of length of pipe sections and bends
- In conclusion, it is commonly and successfully used in the UK gas industry for joint encapsulation work, mainly due to the absence of alternative techniques.

Sonic Gas Research Institute (GRI) Tool

Houston Advanced Research Centre (HARC) and Folsom Research carried out a number of tests with different transducers on the Sonic GRI tool.

The major difficulty was to find one transducer that could provide good signals under two opposite extreme conditions. Sonic waves transmit well through dense wet soil, but not through dry materials. The problems encountered were:

- Achieving a good coupling with the soil surface to transmit the sound waves efficiently into the soil and then to retrieve them to receive a visual image.
- Identifying transducers with the appropriate frequency range to 'look' down 3 ft 3 inches to 6 ft 6 inches (1-2 metres). Low frequency transducers have been used effectively to visualise deep penetration wells.
- One transducer is not suitable and an array of transducers would be better.

According to correspondence with HARC in July 2003, this work is no longer an area of focus and was discontinued some time ago. Use of the technique can be summarised as follows:

- There were no benefits identified
- The major limitation was that the system cannot cater for different soil conditions
- In conclusion, the project was abandoned due to operational limitations. No known further research.

Conductivity Assessment

The electrical resistivity/conductivity* of a substance is a measure of the ease with which an electrical current can be made to flow through it. In general the conductivity is electrolytic and takes place through the moisture filled pores and passages, which are contained within the insulating matrix, which in this case would be the soil or other subsurface materials.

A number of factors including, porosity i.e. the shape, size and number of pores, size and shape of interconnecting passages, the extent to which pores are filled by water and other factors influence the value of the conductivity measured. A typical 3 and 2 dimensional output is displayed in Figure 5.6.

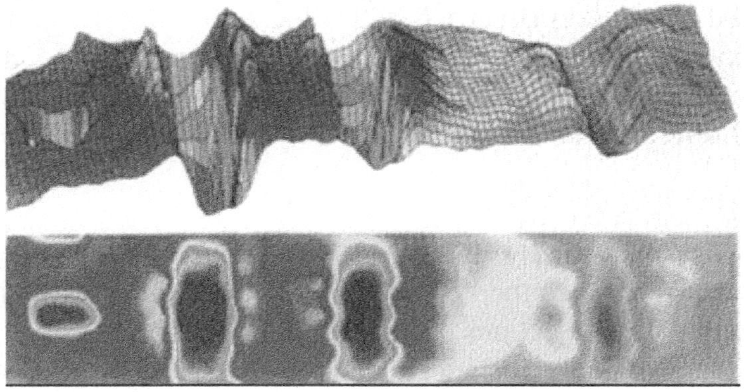

Courtesy Geonics

Figure 5.6 Typical contoured conductivity map showing range of values measured

* Conductivity is inversely proportional to resistivity

Equipment is available which is able to measure conductivity using an induced magnetic field technique. The Geonics EM38 tool, Figure 5.7, was tested on open ground on an industrial estate with no details of any buried apparatus.

Courtesy Geonics

Figure 5.7 Geonics EM38 conductivity measurement tool

The results showed that there were three identifiable conduits, probably gas, water and electricity supplies to the factory units. The conduits were tracked a short distance in order to attempt to identify joints. No significant change was noted at any particular position although the signal varied possibly due to depth and ground moisture.

The Geonics EM41 conductivity tool, Figure 5.8, was tested courtesy of Northumbrian Water Limited at Whittledene water treatment works where the position of buried pipes was known. The purpose was to attempt to identify joints and known wet areas.

Courtesy Geonics

Figure 5.8 Geonics EM41 conductivity tool at Whittledene

There are two components of the induced magnetic field measured by the EM41. The first is the quadrature-phase component, which gives the ground conductivity measurement in milliSiemens/metre (mS/m). The second is the inphase component used primarily in the EM41 for calibration purposes. The inphase measurements are the ratio of the induced secondary magnetic field to the primary magnetic field in parts per thousand (ppt).

It is always advisable, when surveying for buried metallic objects, to record both the inphase and the quadrature-phase components. While the inphase in general, is a better detector of metallic objects, the quadrature-phase is more sensitive to long, extended targets (e.g. pipelines) which are, at least partially, in electrical contact with the ground.

The tool was used initially to locate a 36-inch (900 mm) cast iron main buried at a depth of about 4 ft 11 inches (1.5 metres). In the four tests undertaken, the device was tried with the dipole held parallel to and then normal to the pipe. Both positions readily located the pipe as shown. Once located, another part of the pipeline was followed in order to traverse a number of joints (tests 3 and 4). There were relatively small changes in signal strength along the pipeline, and the device was therefore unable to detect any joints. Signal strength however does vary with changes in ground conditions as shown in Figure 5.9.

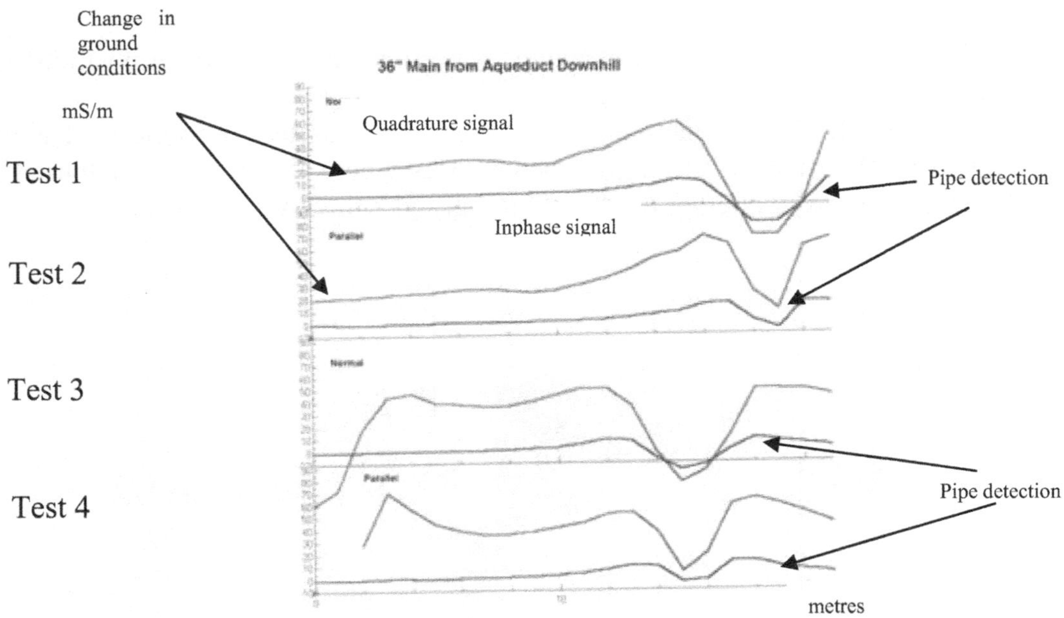

Courtesy Geonics

Figure 5.9 36 inch cast iron main location at different positions

Three other tests were carried out to assess any other potential the device may have. In test 1, a 98 ft (30 metre) traverse referred to across the 'pond', parallel to but away from a DN1200 ductile iron pipe. The results showed some gradual change of signal from one sensor probably due to slightly drier ground conditions. In test 2, a similar traverse away from the DN 1200 pipe showed the influence of the pipe initially, then the presence of a 30 inch (760 mm) cast iron pipe. Although this traverse crossed the 'pond' with a known wet area, this was not apparently detected by the device. In test 3, another traverse clearly located the three outlet mains on both sensors but no other feature. These results are shown in Figure 5.10.

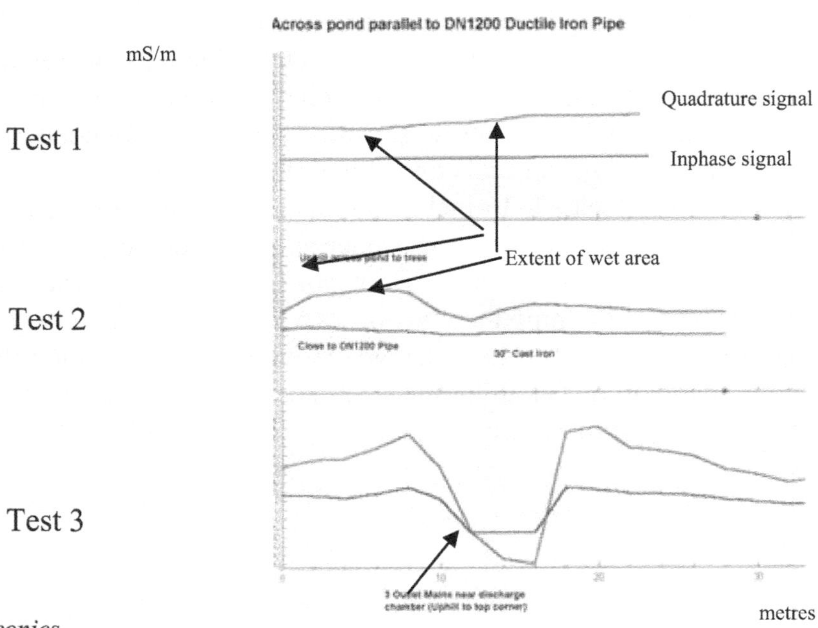

Courtesy Geonics

Figure 5.10 Results from three traverses

The use of the technique can be summarised as follows:

- The benefits of the system are that it can identify pipe and changes in pipe depth and ground conditions
- Limitations are that the EM41 is awkward to use except in open safe areas, and that it was unable to detect pipe joints
- In conclusion, the system is unable to detect pipe joints and no known further research is currently being undertaken

NON INTRUSIVE – ON PIPE

Guided Wave Technique – Ferrous pipe

Long range ultrasonic methods use so called 'guided waves'. These are similar to the Lamb waves, which are generated in plates and in pipe walls but are necessarily of much lower frequency than those used for normal ultrasonic tests used to determine material thickness. Typically frequencies around 50kHz are used compared to around 5MHz for conventional thickness testing. These waves have the property that they can travel many metres along continuous and homogeneous metallic pipes with minimal attenuation and therefore offer the potential of testing large areas of pipe from a single point using a pulse echo transducer bracelet wrapped around the pipe. Any changes in thickness of the pipe, and features such as welds and joints, cause reflections, which are detected by the transducer.

The device has also been successfully tested on single spool cast iron socket and spigot pipes as shown in Figure 5.11 with the output illustrated in Figure 5.12. Here the range was limited to one mechanical joint.

Courtesy Guided Ultrasonics Ltd.

Figure 5.11 Bracelet on cast iron pipe

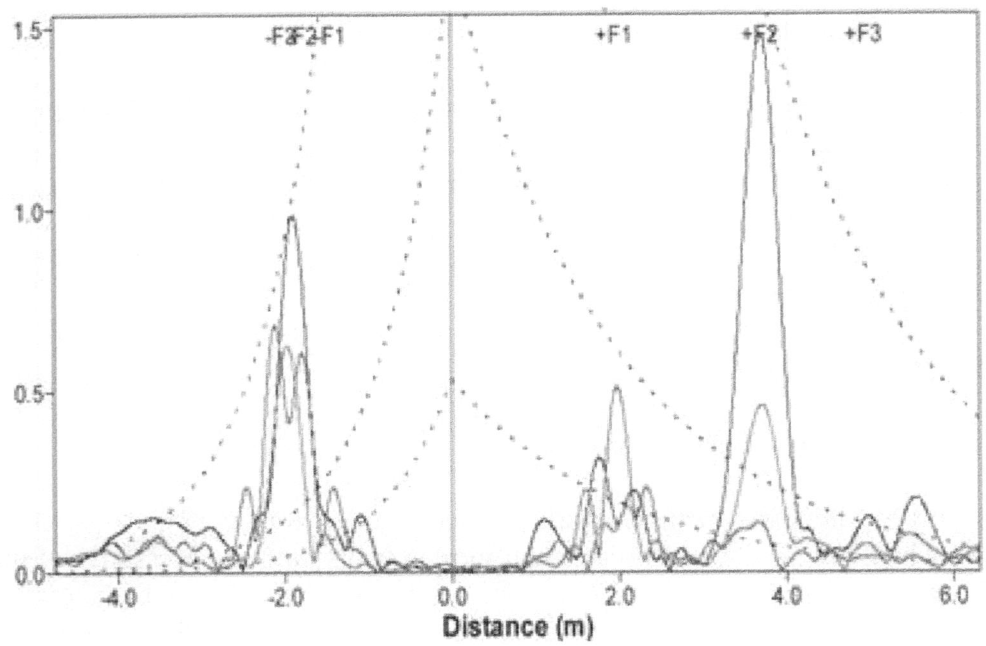

Courtesy Guided Ultrasonics Ltd.

Figure 5.12 Guided wave output – cast iron joints

The use of the technique in this application can be summarised as follows:

- Benefits - Can identify all butt-welded joint positions accurately up to 80 metres pipe length in both directions. Non-intrusive.
- Limitations - Single joint only for mechanical joints. Requires excavation for buried pipe
- Conclusion - Limited application for joint detection. No known usage for joint detection only.

Guided Wave Technique – Plastic Pipe

Limited tests have been carried out using guided waves on polyethylene (PE) pipes. Further development would be needed to address the signal attenuation problem and to detect joints. It is considered that a joint could be identified and its condition assessed, particularly as a quality control tool, if there was a demand. The use of the tool in this application can be summarised as follows:

- The benefit is that the tool is non-intrusive
- The major limitation is poor transmission of Lamb waves due to signal attenuation in PE. Requires excavation for buried pipe
- There is no known further research being undertaken for use on PE pipes.

Current Grading System – NYSEG

This technique has been developed by the New York State Energy and Gas company and has had limited success on gas pipes. Induced currents are transmitted into the pipe, which meets greater resistance at joints and leaks into the surrounding soils. These are detected and measured at the surface. It was considered by NYSEG that the conductivity of water might increase the signal strength transmitted and improve the quality of the output. The use of this technique in this application can be summarised as follows:

- Benefits – Can detect many joints from one transmission point.
- Limitations – Mixed success due to different ground conditions. Requires excavation or pipe access
- Conclusion - Limited success in US Gas Industry. Some potential in Water industry, but requires further development

INTRUSIVE (IN PIPE)

Caliper Pig

Originally designed to provide information on pipeline defects and features related to pipeline geographical location and orientation, the caliper pig, Figure 5.13, could be adapted to locate joints.

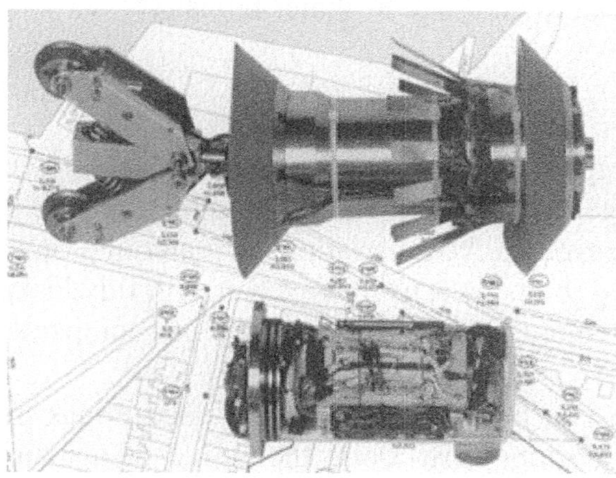

Courtesy Advanced Engineering Solutions Ltd.

Figure 5.13 Caliper pig

Figure 5.14 shows a feature in which all of the calliper sensors have dropped into a recess of maximum depth 35mm and width 65mm. It is considered likely that this feature is a flanged joint or a socket and spigot joint.

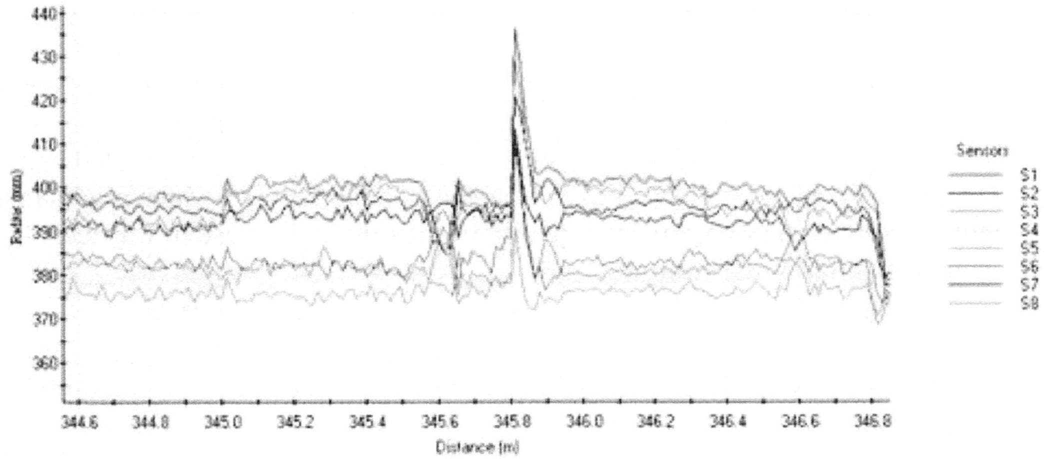

Courtesy Advanced Engineering Solutions Ltd

Figure 5.14 Typical output showing joint detection

The use of this system for this application can be summarised as follows:

- The benefits are that all joints are located on a long run
- The limitations are that the system requires a launch and receive facility
- In conclusion, the system is commonly used in the Gas and Oil Industries and it can detect joints. Unknown usage in Water Industry

Hall Effect Sensors

The Hall Effect principle is to locally saturate the pipe wall with magnetic flux. This results in excess flux leakage. Hall Effect sensors measure this base flux leakage density and hence detects any changes resulting from variations in material properties or dimensions. Also, they are capable of detecting joints by sensing the change in flux leakage at the air gap. For example, this technique is widely used on live mains by gas companies to detect lead/yarn joints for anaerobic spraying.

Anaerobic spraying is a joint repair technique used in gas mains. These repair systems typically involve rodding a spray head along the inside of a main under live conditions and spraying anaerobic sealant into the joint gaps of spigot and socket joints. Joints are detected using the Hall Effect sensor, to allow positioning of the anaerobic spray head. The principle is illustrated in Figure 5.15 with an output identifying a number of joint traverses in Figure 5.16.

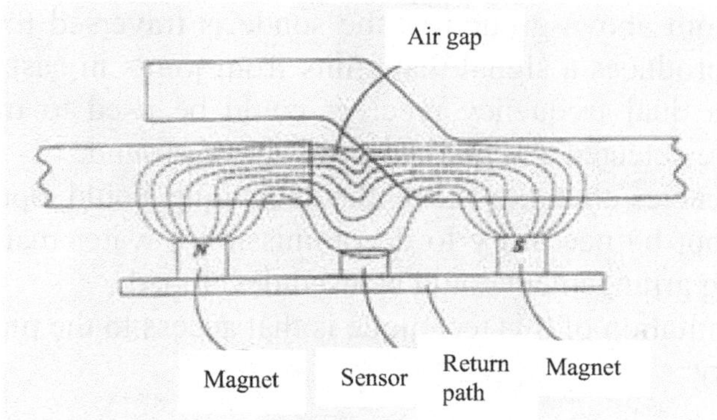

Figure 5.15 Hall Effect principle

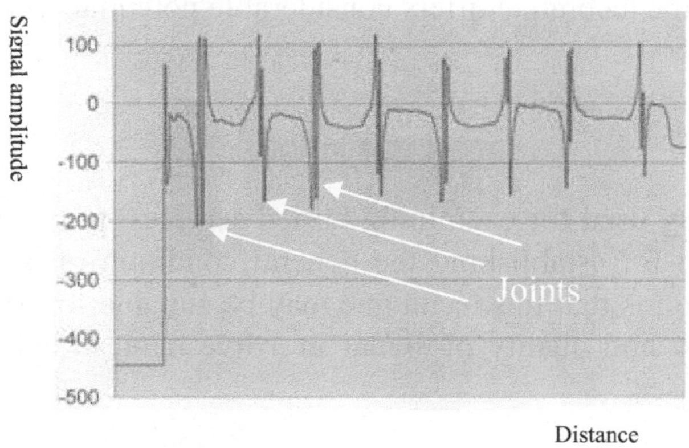

Courtesy Advanced Engineering Solutions Ltd
Figure 5.16 Output - joint traverses

This technique is currently being developed for the positioning of a conceptual joint sealing tool designed for use in water mains under live conditions. The use of the tool for this application can be summarised as follows:

- The benefits are that the technique is simple and effective and that it can be installed on other devices
- The limitations are that small, corroded and/or debris filled joints can be overlooked or difficult to identify
- In conclusion, the technique is commonly used in UK Gas industry for detection of joints during anaerobic spraying

Sonde and Receiver Technique (Dual Frequency Radio)

The sonde and receiver technique is commonly used to trace the route of non-metallic pipes, sewers, drains and ducts. The technique involves the use of a sonde (transmitter) that is rodded through a buried pipeline and tracked from above ground using an appropriate receiver.

In theory, an enhanced version of this technique could be used for the location of mechanical joints in metallic pipes. For this purpose, a dual frequency sonde and receiver system would be required. A low frequency sonde (say 20 Hz) emits a signal that can penetrate the wall of metallic pipes and can be picked up by a receiver above ground. This allows the route of the pipe to be tracked from above ground as the sonde is traversed through it. A high frequency sonde (say 32 kHz) produces a signal that emits from joints in cast iron pipes, but not from the pipe barrel. Thus, a dual frequency receiver could be used to track the pipe and allow the position of joints to be detected and marked from above ground.

At the frequencies that are used, the technique could operate in water filled mains. Therefore, it would not be necessary to decommission a water main, as long as an appropriate under pressure sealing arrangement could be used/developed.

A practical limitation of this technique is that access to the pipe is required for insertion of the sonde. In summary:

- The major benefit of this system is that all joints are located on a long run
- The limitation is that the system requires an in-pipe launch facility
- In conclusion, the technique offers considerable potential, but no commercial supplier has been found.

CCTV

CCTV is extensively used for examining buried services, principally for the detection of pipe wall failures and also for establishing the general condition of sewers, Figure 5.17. The current consensus of opinion is that this technique may be suitable for joint detection in non-live water mains but the clarity and quality of signal in a live main is uncertain, especially as the diameter of the main increases.

Courtesy Northumbrian Water Ltd

Figure 5.17 Launch of CCTV camera

The use of the tool for this application can be summarised as follows:

- The benefits are that the tool can identify joints
- The limitations are the clarity of water/turbulence
- In conclusion, the system has potential for use in the water industry. Some usage to date in raw water in UK.

Magnetic Flux Leakage Pig

Magnetic flux leakage tools (MFL), which can be designed to scan both inside and outside surfaces of a pipeline employ magnets placed next to the pipe wall under investigation.

When a magnet is placed next to a ferrous pipe wall most of the magnetic flux lines pass through the pipe wall. The pipe is the preferred path for the flux completing the circuit between the north and south magnetic poles. Leakage of magnetic flux will occur where the pipe wall thickness has been reduced, possibly due to corrosion or some other form of damage. A suitable sensor placed in this location, measures the level of flux leakage. With the appropriate calibration the size and shape of a defect can be determined from this flux leakage measurement.

Whilst primarily intended for the detection of corrosion and damage this type of tool, Figure 5.18 can detect joints in metallic pipes, and could, with additional equipment and technology, assess the condition of joints which have been detected.

Courtesy Advanced Engineering Solutions Ltd

Figure 5.18 MFL inspection pig

Operationally, there are limits on size and wall thickness. The use of the system for this application can be summarised as follows:

- The main benefit is that all joints are located on a long run
- The limitations are that water pipe may require prior cleaning and will require a launch and receive facility
- In conclusion, the system is commonly used within the Gas and Oil Industries and can detect joints. Unknown usage in Water Industry

Remote Field Eddy Current Device

The remote field eddy current (RFEC) device consists of an exciter coil and a detector coil placed at a predetermined distance apart. The principle of operation is illustrated in Figure 5.19. The device is passed through the pipe being inspected and the exciter coil is fed with a low frequency alternating current. This exciter current generates an electromagnetic field around the exciter coil. The energy of the field spreads out in an axial direction inside the pipe as well as into the pipe wall. The eddy currents, which are induced in the pipe wall, generate a secondary field, which can be measured by the detector module and interpreted into wall loss and other pipeline defects.

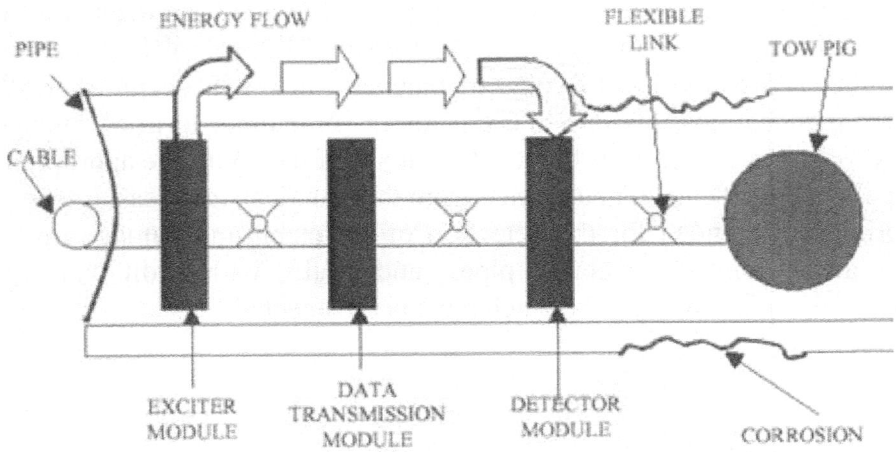

Courtesy Advanced Engineering Solutions Ltd
Figure 5.19 RFEC principle

The use of this technique in this application can be summarised as follows:

- The benefit is that all joints are located on a long run
- The limitation is that the water pipe may require prior cleaning and will require a launch and receive facility
- In conclusion, the system has potential for use in water industry. Some usage in the Canadian Water Industry and the Australian Gas Industry already

Ultrasonics

A number of in-pipe inspection pigs use ultrasound to detect metal loss and cracks in oil and gas pipelines. The use of this technology may allow the detection of mechanical joints in water mains. It is also considered that using a combination of sensors employing direct ultrasound pulses radially into the pipe wall and also at an angle to the pipe wall, could allow different joint types to be distinguished. As these tools are not generally available for the inspection of cast iron water pipelines, any evaluation of their potential for the detection of push-fit joints could not be evaluated within this project. However, the use of internal ultrasonic systems for joint detection and condition assessment is considered in more detail in Chapter 6. The experimental work undertaken suggested that ultrasonics could be used to detect joints.

CHAPTER 6
POTENTIAL CONDITION ASSESSMENT TECHNIQUES FOR JOINTS WITH NON-DETECTABLE LEAKAGE

EXTERNAL TECHNIQUES

Pipe Surveying

Surveying equipment is a possible means of either measuring above ground pipe movement or monitoring for ground movement. This equipment can provide data on angular displacement in both the vertical and horizontal planes. Also, it could provide a measure of total 'pull out' over a displaced length.

Lord Technical Hire were approached to carry out a survey at Heights Quarry, courtesy of Northumbrian Water Limited (NWL). The pipe at this site is a buried 30 inch (760 mm) diameter cast iron socket and spigot pipeline with lead yarn joints laid in 1900 had been subjected to ground movement due to some local mining collapse. During the course of some repairs, the pipeline was partially exposed over a 490 foot (150 metre) length and so provided an ideal site to assess the technique.

The centre line/crown of the pipe was marked with a small dot off either side of each joint using a custom-made (sprit) level, Figure 6.1. These would be the location points for the survey prism/pole.

Courtesy Northumbrian Water Ltd.

Figure 6.1 Marking the pipe centre line

The laser tripod was set up using a (spirit) level. Reference points were selected to define the x,y,z co-ordinates and for any subsequent monitoring. The prism pole is held as still and vertical as possible in the direction of the laser, although some movement and deviation is acceptable. A self-centring prism is used which caters for small movement of the pole and should always line up with the laser signal. A TS315C Laser tool was used, Figure 6.2.

Courtesy Advanced Engineering Solutions Ltd.

Figure 6.2 Surveying at Heights Quarry

The laser tool is pointed towards the prism, locked in position and fine focused by grub wheels; with experience this takes about 10 seconds. The laser is then activated and the x, y, and z co-ordinates are recorded and displayed. The process was repeated and 48 sets of readings were collected in about 40 minutes. The data were saved onto an Autocad file for subsequent examination. The software is not set up for this specific type of interrogation of comparing angular alignments of adjacent pipes. The figures presented in Table 6.1 are derived using Autocad to define relative angular positions.

The results show that the greatest angular deflection between spools occurred at pipe joints 19 and 20 with 2 ° 4' and 3 ° 23' respectively. These are equivalent to +/- axial movement of 0.56 inch (14 mm) and 0.75 inch (20 mm) respectively, shown in Figure 6.3.

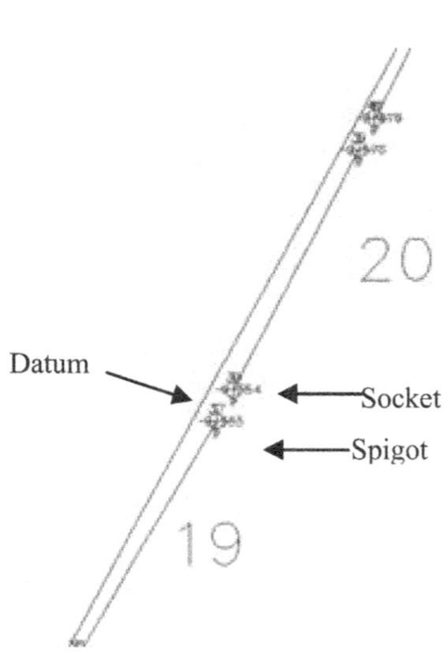

Figure 6.3 Pipes 19 and 20 referenced relative to datum

Table 6.1
Heights Quarry survey results

Pipe spool	Measurement points	Angular deflection, deg and mins	Reference datum
1	1-2	1^0 26’	From Datum Line
2	3-4	0° 17’	Spool 1
3	5-6	0° 70’	Spool 2
4	7-8	0° 36’	Spool 3
5	9-10	1° 19’	Spool 4
6	11-12	0^o16’	Spool 5
7	13-14	0° 5’	Spool 6
8	15-16	0^0 47’	Spool 7
9	16-17	0° 01	Spool 8
10	18-19	0° 58’	Spool 9
11	20-21	0° 28’	Spool 10
12	22-23	1° 46’	Spool 11
13	24-25	0° 12’	Spool 12
14	26-27	1° 17’	Spool 13
15	28-29	1° 8’	Spool 14
16	30-31	0° 14’	Spool 15
17	32-33	0° 52’	Spool 16
18	34-35	0° 11’	Spool 17
19	36-37	2° 4’	Spool 18
20	38-39	3° 23’	Spool 19
21	40-41-42	0° 34’	Spool 20
22	43-44	1° 8’	Spool 21
23	45-46	0° 35’	Spool 22
24	47-48	0° 31’	Spool 23

The joint between spools 19 and 20 does exhibit a weep from the crown as shown in Figure 6.4.

Courtesy Northumbrian Water Ltd.

Figure 6.4 Weep at joint on pipes 19 and 20

From the height readings illustrated in Figure 6.5, the first spool has been displaced upward by about 4 inches (100mm) whilst spools 2 and 3 are displaced downward relative to spool 1, back to the original datum. Pipe spool 4 is displaced downward by 6 inches (150mm) relative to original datum. Thereafter, there is a gradual further 4 inches (100mm) downward displacement up to spool 10 and then a gradual relative upward recovery up to spool 24, then a sharp return to the original datum. This shows over most of the exposed length, the pipeline has moved downwards relative to the end positions.

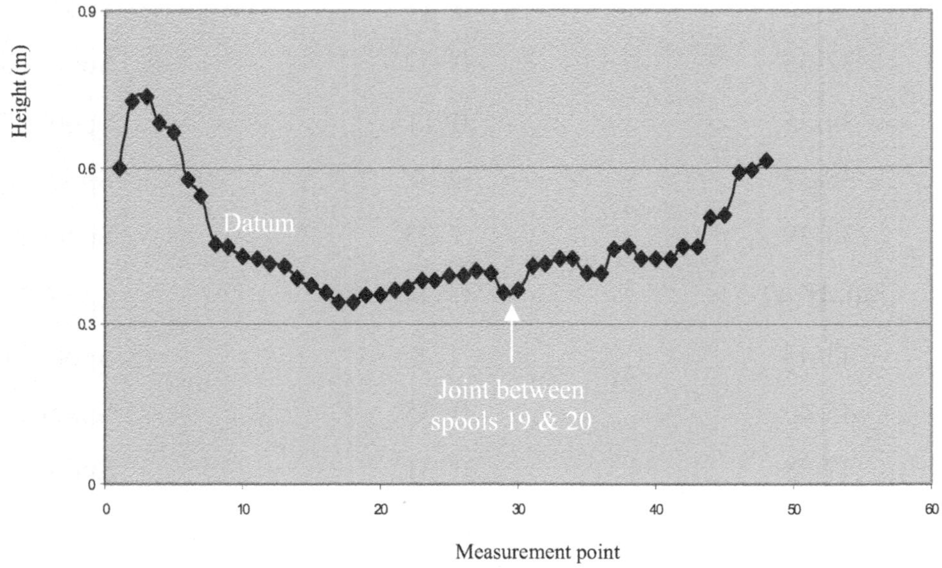

Figure 6.5 Height variations

Permission was granted by Northumbrian Water Limited to conduct a survey on a pipeline at East Hartford, Figure 6.6.

Courtesy Advanced Engineering Solutions Ltd.
Figure 6.6 East Hartford survey

This twin 18 inch (450 mm) diameter cast iron pipeline has suffered a number of leaks along an above ground section leading up to a bridge crossing. The pipeline is known to run through a mining area and it is highly probable that mining subsidence resulted in vertical movement of the pipes. The extent of leakage and number of leaking lead/yarn joints is unknown but was sufficient for the farmer to complain and remedial action was taken by Northumbrian Water. Split collars were installed at every joint tightened up to the spigot face and then injected with a silicon type sealant, Figure 6.7.

Courtesy Northumbrian Water Ltd.
Figure 6.7 Repair installation

It is not considered likely that any realignment of the joint was achieved in this process. Hence the resultant relative angular displacement between spools following the repair is probably unchanged. The results from the survey, Figure 6.8 clearly shows no horizontal (sideways) displacement of the pipeline.

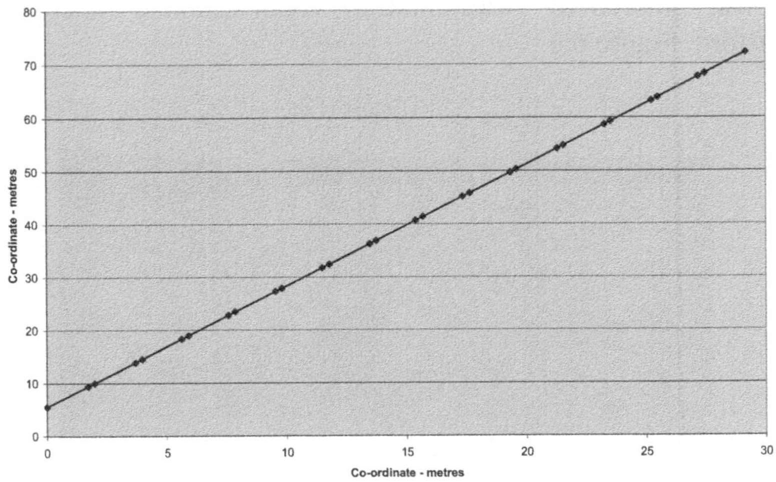

Figure 6.8 Horizontal co-ordinates at East Hartford

The vertical displacement results however do show small variations between adjacent pipes, Table 6.2. Without knowing the original as laid angles, it is not possible to attribute the variations solely to subsidence. However, the pipe was not leaking as new and subsequent movement did result in failure of the seal at some of the joints.

The maximum vertical angular displacement was 0.5 degrees, and over 15 spools, average 0.1 degrees. Although these figures are small, if these are the result of ground movement and not the original line then they need to be put into context. The maximum recorded angular deflection of 0.5 degrees, is equivalent to a 1.19 inch (29 mm) deviation at the point where the pipe enters the next joint and total rotation within the joint of 0.19 inch (4 mm). This could imply that even such a small rotation could lead to leakage.

Table 6.2
Vertical displacement at East Hartford

Pipe Reference	Vertical angular alignment from datum
1-2	*0.02*
2-3	-0.12
3-4	0.23
4-5	0.03
5-6	0.28
6-7	0.34
7-8	-0.19
8-9	0.22
9-10	-0.31
10-11	0.34
11-12	-0.51
12-13	0.25
13-14	-0.02
14-15	0.09
15-16	-0.20

The ground movement at Heights Quarry was considered to be sudden whereas the movement at East Hartford appears to have been gradual, since there are no subsidence pits in evidence which are typically associated with sudden collapse. The use of this technique for this application can be summarised as follows:

- The benefits of the system are that it is simple and effective. It is ideal for monitoring above ground pipes and areas of suspected ground movement with buried pipes
- The limitation are that with most of the pipes, there is no as-laid data for comparison and how the results can be interpreted to provide a view on joint condition
- In conclusion, the technique offers the potential to monitor areas with ground movement

Ultrasonic Tests

An experiment was set up with The Welding Institute (TWI) North (UK), to evaluate the potential of using ultrasonics across a bolted flange to assess joint condition. TWI are very experienced in the use of ultrasonic transducers as part of their everyday work.

The objectives of the experiment were firstly to establish whether a signal could be transmitted across a gasket and secondly could any meaningful interpretation be made of that signal. A 6-inch (150 mm) diameter flat faced 4 bolt-hole steel flange was used, Figure 6.9.

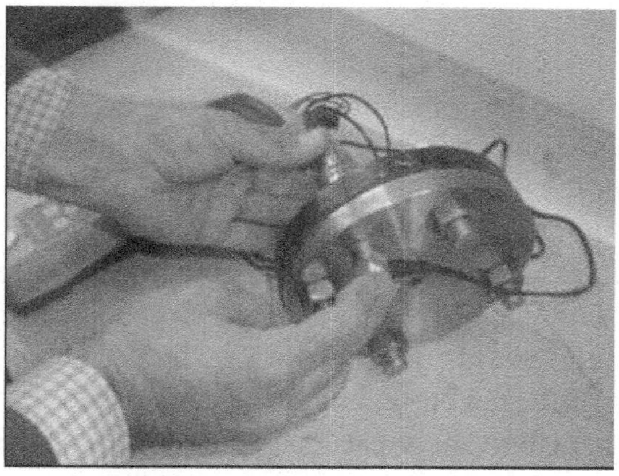

Courtesy Advanced Engineering Solutions Ltd.

Figure 6.9 Ultrasonic experiment on bolted flange

Two single crystal 0.38 inch (10 mm) diameter 4 MHz compression probes were used as transmitter and receiver. Oil based couplant was used as standard practice, rather than water based, to prevent rust.

Test 1: No Gasket

The bolts were tightened without any gasket. As expected by TWI, no signal was transmitted since there would be a small air gap at the flange interface since the faces had not been specially prepared and polished. Apparently a gap > 0.00001 mm will prevent transmission. The transmitter probe was changed back to the pulse-echo mode and a normal signal was displayed, confirming that the probe was working correctly.

Test 2: A Dry 3mm Thick Rubber Gasket

At three positions between bolts, a transmitted mV signal of about 2 units of amplitude was displayed. A high gain was needed, but TWI confirmed that the signals were genuine and not just noise.

At the fourth position, a transmitted signal of 8 units of amplitude was displayed, which varied slightly as the probe was moved closer to the edge of the flange and the bolts. This position was marked on the flange. It was considered that perhaps this may have been due to different tightness, however the tightness of all bolts was judged to be similar. The reason for the better transmission was attributed to a local area of the gasket, which had been tested with couplant and although it had been cleaned up, there was a definite residual mark and presumably some increase in transmissibility.

Test 3: Part Dry, Part Wet

Two of the opposing quadrants were left dry. One remaining quadrant of the gasket was partially wetted with water on both sides and the other fully wetted with water on both sides. The signal output was two divisions of amplitude again at the dry quadrants and slightly over 10 at the fully wet quadrant. The dry portion of the other quadrant registered 2 rising to 5 over the wet section.

Test 4: Fibrous Type Gasket

As expected, there was no transmitted signal as there was air trapped in the gasket structure. Overall, this experiment of four tests was successful since it was not anticipated that there would be any meaningful transmitted output. Although this was just one series of simple tests, it has shown that a sound joint will generate a detectable output and more importantly, differences in gasket interface condition can be detected and the technology could be applied. Whether any quantifiable output can be measured requires further investigation and what the limitations of the system are would also need to be addressed. However, the technique is considered worth pursuing further. The technique was also shown to work on a cast iron/steel bolted flange joint similar to the one shown at East Hartford, Figure 6.10.

Courtesy Northumbrian Water Ltd.

Figure 6.10 Cast iron/steel bolted flange

Note that this technique can only be used on old full flat faced flanged pipe joints. The use of this technique for this application can be summarised as follows:

- The major benefit is that the system is simple and effective
- The main limitation is that it can only be used on full diameter flat faced flanges only.
- In conclusion, the technique can identify changes in gasket condition, but requires further development

Guided Wave

The principles of guided wave technology were described in the section on joint detection in Chapter 5. In order to assess whether guided wave could detect variations in welded joints, five sections of 2-inch (50 mm) diameter steel pipe were especially prepared using MIG and TIG welding equipment. One weld was carried out correctly, whilst for the other joint, poor preparation and weld procedures were used. Guided Ultrasonics Limited (GUL) were invited to conduct an experiment on the pipe section shown in Figure 6.11, to determine whether these features could be identified.

Courtesy of Guided Ultrasonics Ltd.

Figure 6.11 Guided wave experiment

The results and GUL comments are incorporated in Figure 6.12. This test has shown that guided wave can detect good and bad welds on steel pipe. In normal practice, where the technique is being used to detect corrosion/defects along a continuous pipeline up to 200 ft (60 metres), the form of the transmitted signal could be modified to provide better interpretation at welds whilst forfeiting distance. Nevertheless, this limited test has shown that where there is concern about welded joint construction and/or condition, then this technique could provide useful data and is worthy of future consideration.

A further field test was undertaken to determine whether the technique could provide any useful information on cast iron socket and spigot jointed pipe.

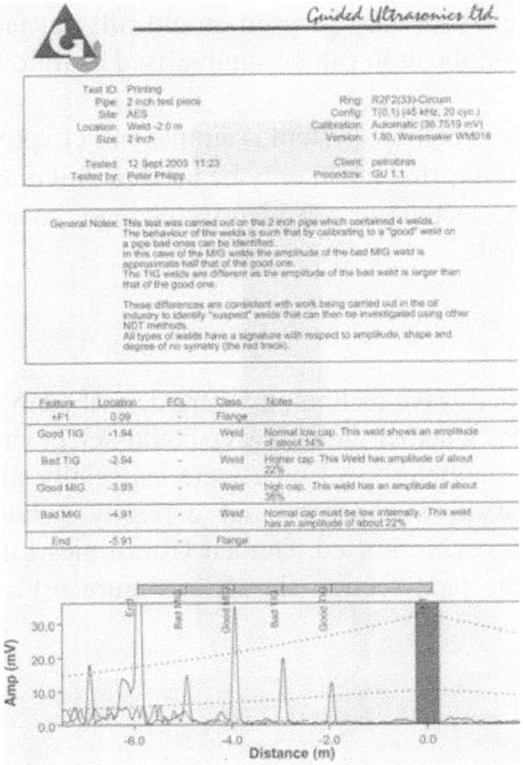

Guided Ultrasonics Ltd.

Test ID: Printing
Pipe: 2 inch test piece
Site: AES
Location: Weld -2.0 m
Size: 2 inch

Tested: 12 Sept 2003 11:23
Tested by: Peter Phipp

Ring: R2F2(33)-Circum
Config: T(0,1) (45 kHz, 20 cyc.)
Calibration: Automatic (36.7519 mV)
Version: 1.80, Wavemaker WM018

Client: petrobras
Procedure: GU 1.1

General Notes: This test was carried out on the 2 inch pipe which contained 4 welds.
The behaviour of the welds is such that by calibrating to a "good" weld on a pipe bad ones can be identified.
In this case of the MIG welds the amplitude of the bad MIG weld is approximate half that of the good one.
The TIG welds are different as the amplitude of the bad weld is larger than that of the good one.

These differences are consistent with work being carried out in the oil industry to identify "suspect" welds that can then be investigated using other NDT methods.
All types of welds have a signature with respect to amplitude, shape and degree of no symetry (the red trace).

Feature	Location	ECL	Class	Notes
+F1	0.09	-	Flange	
Good TIG	-1.94	-	Weld	Normal low cap. This weld shows an amplitude of about 14%
Bad TIG	-2.94	-	Weld	Higher cap. This Weld has amplitude of about 22%
Good MIG	-3.93	-	Weld	high cap. This weld has an amplitude of about 36%
Bad MIG	-4.91	-	Weld	Normal cap must be low internally. This weld has an amplitude of about 22%
End	-5.91	-	Flange	

Courtesy Guided Ultrasonic Ltd.
Figure 6.12 Guided wave results and comments

Joint Condition East Hartford

The purpose of this test was to demonstrate the principle of assessing the socket and spigot ends of a single spool for the detection of corrosion and/or defects in cast iron pipe. An 18 inch (450 mm) diameter pipe at East Hartford was used, Figure 6.13.

Courtesy Guided Ultrasonic Ltd.
Figure 6.13 Guided wave equipment on an 18 inch cast iron pipe at East Hartford

The results and comments from GUL are shown in Figure 6.14.

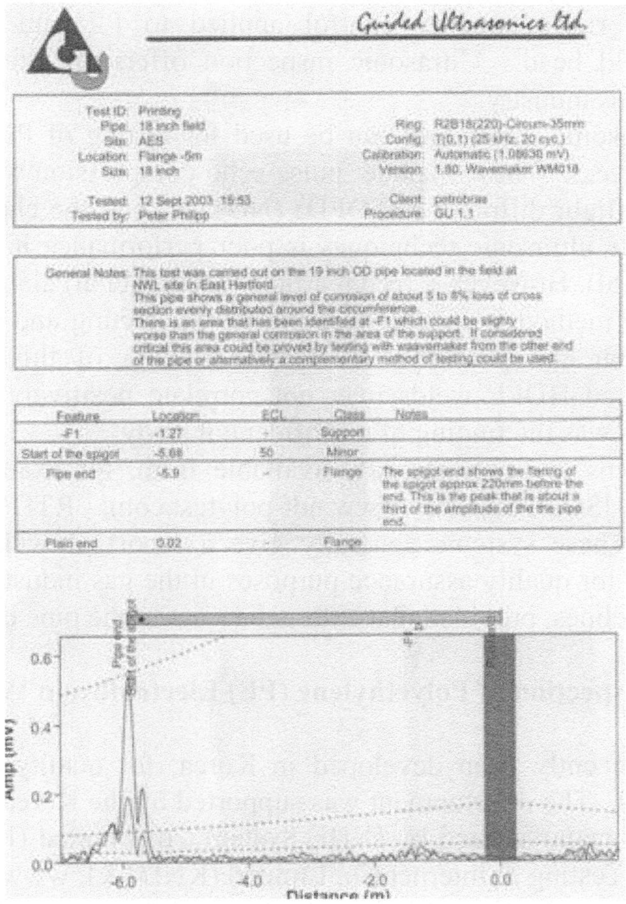

Guided Ultrasonics Ltd.

Test ID: Printing
Pipe: 18 inch field
Site: AES
Location: Flange -5m
Size: 18 inch

Ring: R2B18(220)-Circum-35mm
Config: T(0,1) (25 kHz, 20 cyc.)
Calibration: Automatic (1.08630 mV)
Version: 1.80, Wavemaker WM018

Tested: 12 Sept 2003 15:53
Tested by: Peter Philipp

Client: petrobras
Procedure: GU 1.1

General Notes: This test was carried out on the 19 inch OD pipe located in the field at NWL site in East Hartford.
This pipe shows a general level of corrosion of about 5 to 8% loss of cross section evenly distributed around the circumference.
There is an area that has been identified at -F1 which could be slightly worse than the general corrosion in the area of the support. If considered critical this area could be proved by testing with wavemaker from the other end of the pipe or alternatively a complementary method of testing could be used.

Feature	Location	ECL	Class	Notes
-F1	-1.27	-	Support	
Start of the spigot	-5.68	50	Minor	
Pipe end	-5.9	-	Flange	The spigot end shows the flaring of the spigot approx 220mm before the end. This is the peak that is about a third of the amplitude of the the pipe end.
Plain end	0.02	-	Flange	

Courtesy Guided Ultrasonics Ltd.

Figure 6.14 Guided wave results on socket & spigot pipe

Due to the shape of the socket and to a lesser extent the spigot, interpretation of the results is more difficult at those positions. However, the form of signal could be adapted to suit this form of interrogation if there was a demand.

The tests have shown that guided wave could be utilized to investigate a number of steel welded joints quite readily and could be developed to provide a viable alternative to other inspection techniques. On the other hand, although interrogating data on socket and spigot pipe could be further developed, inspection would be restricted to one pipe spool at a time and is therefore not considered to merit further investigation. The use of this technique for this application can be summarised as follows:

- The benefit of the system is that it can examine several joints from one transmission point on butt welded metallic pipes
- The limitations of the system are that it requires an excavation for access to buried pipes and can only examine one pipe spool for non butt welded pipes
- In conclusion, it can identify variations in weld performance but it requires further research and development

External Ultrasonic Inspection of Polyethylene (PE) Butt Fusion Welds

The common form of quality control applied to PE butt fusion welds is visual examination of the weld bead. Ultrasonic inspection offers an alternative approach that is sometimes used in the gas industry.

A range of ultrasonic techniques can be used for testing of PE butt fusion joints, with differing degrees of success. These include; pulse-echo compression wave, pulse echo creeping wave, tandem, time of flight diffraction (TOFD) (NDTnet) and the chord technique (Giller). A noted limitation of some ultrasonic techniques is poor performance in detecting lack of fusion (cold welds) (Rad'ko et al). However, a recent article (Messer et al) claims that a new application of the ultrasonic TOFD method is capable of accurately detecting and quantifying weld defects including lack of fusion and porosity. A further finding of this article was that visual examination of completed HDPE welds may not correlate positively with weld integrity. It should be noted that this was the finding of a confidential study.

Ultrasonic systems are commercially available from, for example; GASTEC, McJem, www.gastecatcre.com; JSC Polytest, www.ndt-polytest.com; RTD Quality Services Inc. www.rtdquality.com. These systems generally give a report of weld condition on site and although normally used for quality assurance purposes in the gas industry, they could be applied to operational water pipelines, provided that external access to the pipe could be gained.

External Ultrasonic Inspection of Polyethylene (PE) Electrofusion Welds

A system has recently been developed in Korea, for quality assurance testing of PE electrofusion (EF) joints. The development was supported by the Korea Gas Safety Corporation. The system (AIM33) is manufactured by INDE System Co. Limited (Korea) and distributed by Korea Non-Destructive Testing & Internet Co. Limited (KNDT&I, www.kndt.co.kr). The system is currently being marketed to the gas industry. The system uses an ultrasonic array approach. Operating principles and examples of results from experimental work are reported in a conference hosted by Yedlitepe University (Turkey). Figure 6.15 shows the AIM33 equipment.

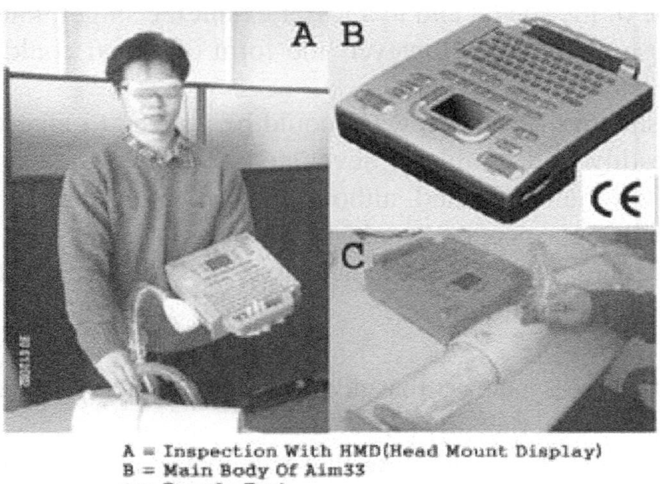

A = Inspection With HMD(Head Mount Display)
B = Main Body Of Aim33
c = Sample Test

Courtesy KNDT&I Co. Ltd.

Figure 6.15 AIM33 equipment

Some of the fusion problems that can be detected by the system are illustrated in Figures 6.16 to 6.19. The figures show cross sections through EF joints, together with the output from ultrasonic scans.

Figure 6.16 shows an example of a good EF joint. The ultrasound scan shows front (top of image) and back wall echoes and clear echoes from the heating wires.

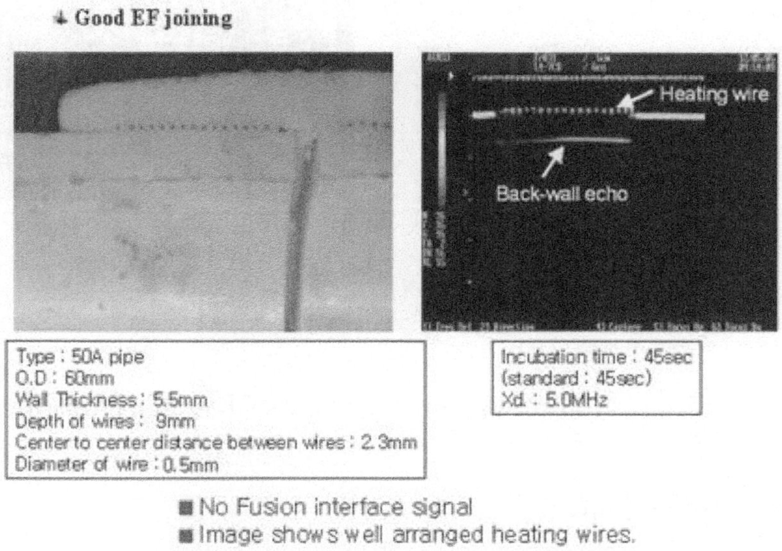

Courtesy KNDT&I Co. Ltd.

Figure 6.16 Good EF joint

Figure 6.17 shows an imperfect EF joint. There is no back wall echo since no fusion has taken place.

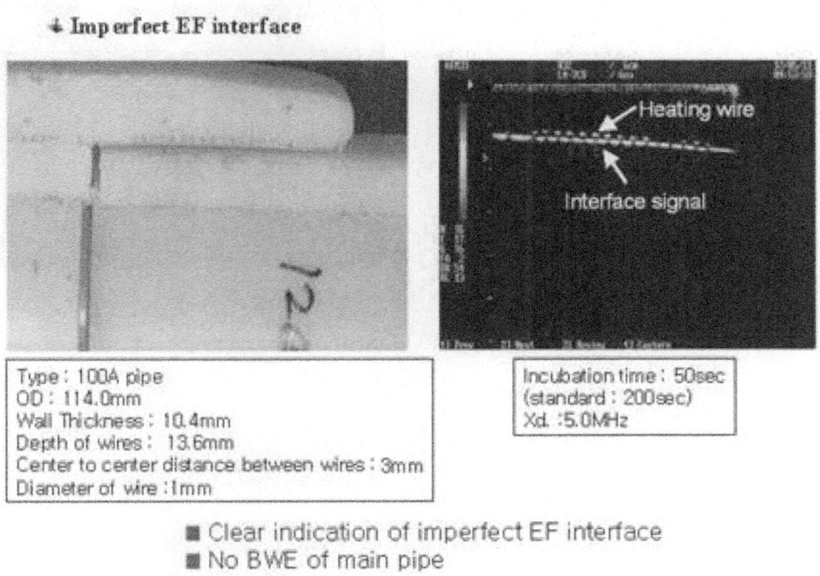

Courtesy KNDT&I Co. Ltd.

Figure 6.17 Imperfect EF joint

Figure 6.18 shows an overheated joint, exhibiting misalignment of the heating wires.

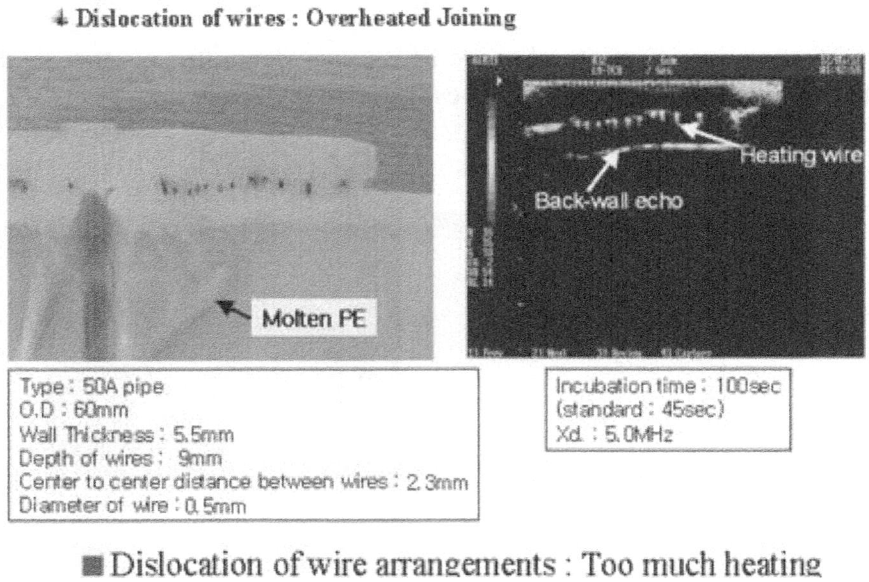

Courtesy KNDT&I Co. Ltd.
Figure 6.18 Overheated EF joint

Figure 6.19 shows the effects of soil inclusion at the fusion interface. The ultrasonic scan shows a clear interface signal between the heating wires, as a result of the soil inclusion.

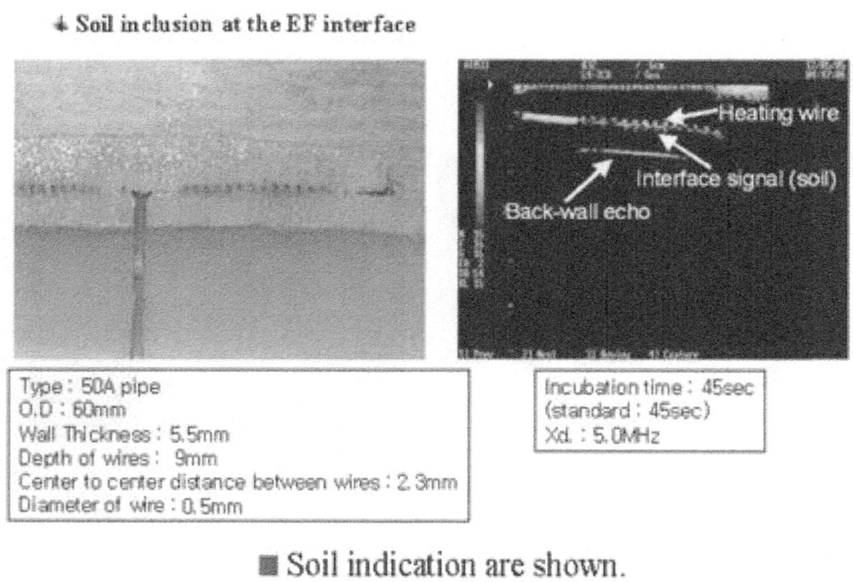

Courtesy KNDT&I Co. Ltd.
Figure 6.19 Soil inclusion at EF interface

The above examples show the capability of AIM 33 for condition assessment of polyethylene EF joints. Although the system is being marketed for quality assurance of newly made joints in gas pipelines, there is no technical reason why the system would not be applicable to quality assurance of new joints and condition assessment of existing joints in water pipelines. In the case of existing buried pipelines, it would be necessary to excavate onto the joint to facilitate the external inspection, but there would be no interruption to water supply. Although the system can detect specific defects, there is currently no real guidance from this or other systems as to the significance of the defects they find in relation to the long term performance of the joint. Therefore, the decision the utility should take e.g. to leave the joint in place or to cut out and remake a new joint, is still difficult.

X-ray External Surface

It has been considered that X-ray equipment could be used to determine joint condition by both external and internal means. X-ray technology is commonly used for the quality control examination of new welds on steel pipelines. It is uncertain whether this could be applied to other types of joints, especially those that have already been in service. It is understood that there are currently no published R&D activities currently being undertaken in this field. The use of this technique for this application can be summarised as follows:

- The main benefit of the system is that it can identify cracks, areas of corrosion and partial leak paths.
- Operationally it is difficult to deploy and requires excavation
- In conclusion, the technique requires further development for this application and current usage is unknown.

Soil Properties

Techniques for assessing soil properties such as resistivity, pH, sulphates etc. are well established, but are also worthy of inclusion as part of a joint condition assessment programme. It may be possible to identify soils that are aggressive to joint components such as bolts, couplers etc. and also to gasket and seal materials exposed to soils or surrounding water.

Variations in the water table will influence the potential for corrosion cell development, changes in ground condition and stability. Similarly dry and/or extreme cold conditions can give rise to local ground movement, which can put extra loads on joints and /or joint components over and above general ground movement. Hence it is recommended that soil properties and associated factors are given due consideration and standard techniques and equipment be used to monitor them. In summary,

- The benefits are that it is simple and low cost to test soils for their aggressiveness to joint components and gasket materials.
- The major limitation is in interpretation of results
- In conclusion, there is some potential use in areas prone to corrosion or gasket degradation. Usage of soil properties for this application is unknown

INTERNAL TECHNIQUES

CCTV/Sonar

The design of TV cameras had advanced sufficiently by 1965 to allow them to be used to internally inspect buried pipelines for evidence of deterioration and the need for rehabilitation. This allowed an increase in the operational life of the pipelines to be achieved by the prompt reporting, targeting and subsequent repair of sections of pipe identified as giving cause for concern following the CCTV inspection.

With the advent of Charged Coupled Devices (CCD) instead of the conventional cathode ray tube in the camera, it was possible to significantly reduce the size of the camera and improve its reliability. The introduction of Pan and Tilt mechanisms and CCD devices increased the versatility of the camera units enabling a wide range of pipe diameters and internal defects to be fully investigated.

The early use of CCTV was to some extent marred by the lack of consistency in describing defects. This caused a number of serious pipeline defects to be overlooked, resulting in a few expensive and very public incidents. These early problems have largely been overcome due to work carried out by the Transport and Road Research Laboratory (TRRL) in the UK, who put forward a Defect Coding System that now forms part of CCTV contract inspection law world-wide. CCTV now looks to be well established as an internal pipeline inspection tool, which could be utilised for the location and subsequent assessment of pipe joints.

Sonar devices, which are generally used in fully surcharged pipelines, have recently been introduced to complement CCTV. In partially surcharged pipelines the two techniques can be used in combination, the CCTV system above the water line and sonar below.

A typical CCTV system was evaluated in an 8 inch diameter pipe test loop. Following the launch, the system traversed along the loop, and several joints were inspected. The versatility of the device with its tilt and pan head were clearly shown.

Whilst joints could be easily detected by both techniques, interpretation of the image in terms of joint condition remains a problem, although interpretation may improve with experience. This is a fairly rapidly moving market and worthy of further consideration for joint detection coupled with some other investigative tool or rehabilitation device. It could also form part of an inspection programme of the pipeline internal condition. The use of these tools in this application can be summarised as follows:

- The main benefit is that the systems allow all round visual inspection and a number of attachments are possible to increase the functionality
- The clarity of the water and the effect of turbulance can be a limitation, along with interpretation of the results
- In conclusion, the system has potential for use in water distribution pipelines, and has currently already been used in raw water pipelines

Caliper Pig

Calipers can be installed on pipeline inspection pigs (as shown in Figure 6.20) to measure pipe diameter and local defects.

Courtesy Advanced Engineering Solutions Ltd.
Figure 6.20 AESL inspection pig with caliper sensors and distance wheel

Calipers are capable of identifying a kicked joint, i.e. displacement misalignment as shown in Figure 6.21, but further development is required to measure angular displacement.

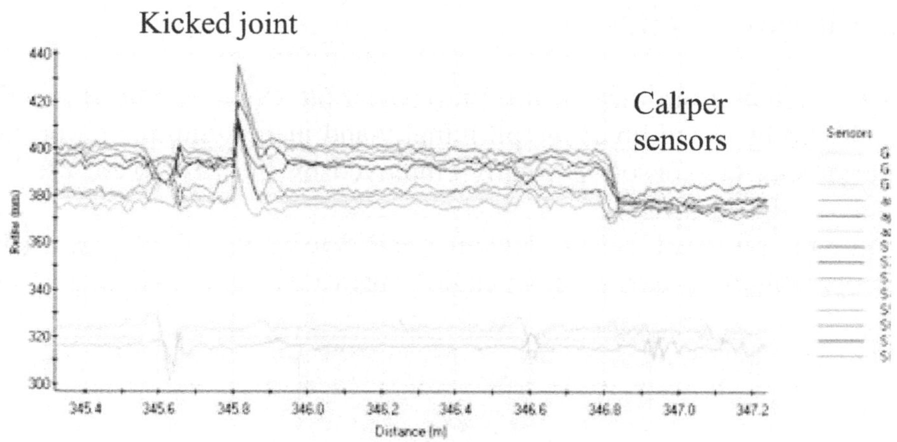

Courtesy Advanced Engineering Solutions Ltd.
Figure 6.21 Joint displacement/misalignment

The use of this technique in this application can be summarised as follows:

- The main benefit is that all joints are inspected on long pipe run
- The main limitations are that the tool requires a launch and receive facility and that there will generally be unknown as laid data to assist interpretation of results for joint condition
- In conclusion, the technique has potential for monitoring, but requires development. It is well established in the Gas and Oil Industries but unknown usage in water

Low Flux Magnetic Sensors

One technique considered for the determination of angular displacement is to measure the joint gap at four quadrants around the pipe circumference using low flux magnetic sensors. The angle across joints could then be determined. Furthermore the joint gap itself could be measured and compared with the specified gap. This could be quite critical in ductile iron pipe where the depth of insertion has a tolerance, which is dependent upon any pre-set angular alignment.

The low flux sensor principle is used to detect and measure corrosion pits/defects in metallic pipes, as installed on the external defect tool (EDT), Figure 6.22 developed by AESL.

Courtesy Advanced Engineering Solutions Ltd.

Figure 6.22 External defect tool (EDT)

Where as the signals from the sensor(s) traverse a pit, the magnetic flux will follow the easiest route, which will be to the lip of the pit initially and then within the pit profile. Hence to demonstrate the principle, the sensors currently employed by the EDT were used to look at a simulated joint gap.

To determine the potential use for detecting and determining joint gap, a laboratory test was developed comprising two plates set at an angle to provide a variable joint gap, Figure 6.23.

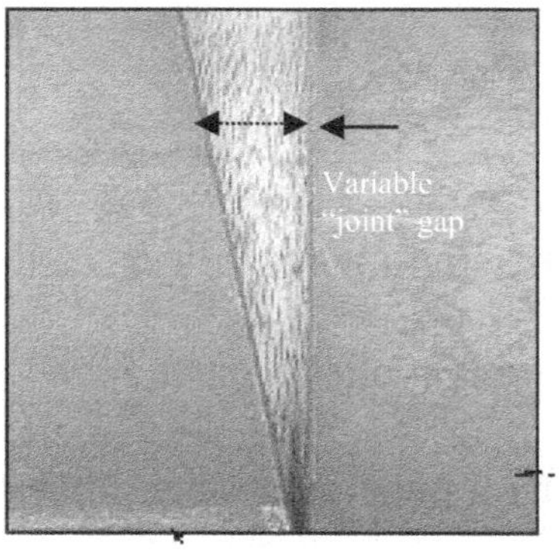

Figure 6.23 Simulated varying joint gap

The tool was passed over the plates from left to right and each sensor responded to the gap corresponding to its position. The signal output results in the form of flux lines are shown in Figure 6.24 and joint gap dimensions in Figure 6.25.

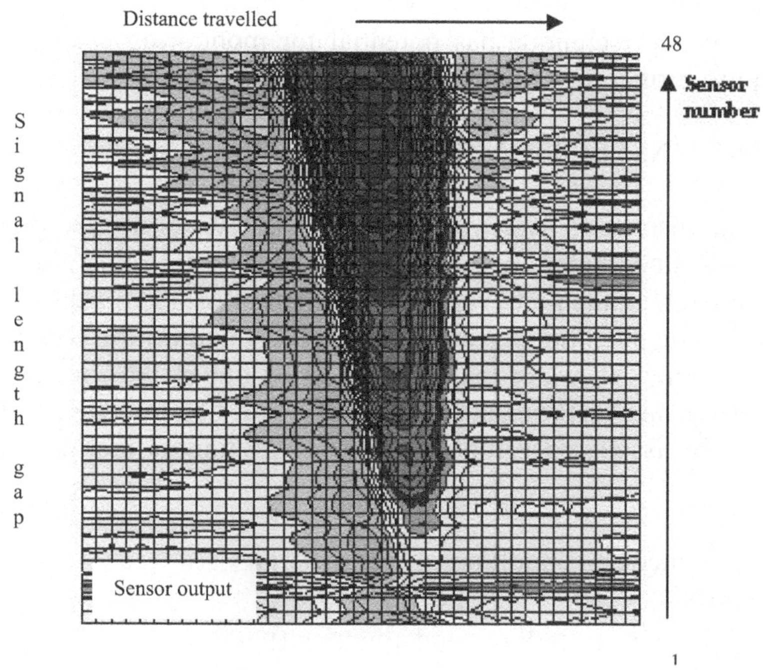

Figure 6.24 Signal output – flux lines

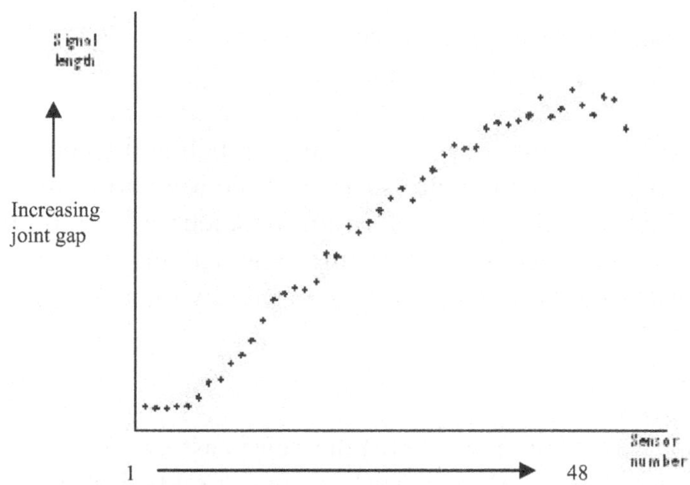

Figure 6.25 Representative gap dimensions

The tests showed that the sensors can detect and measure a gap and could provide a practical technique for joint detection, joint gap measurement and angular displacement. It should be noted that although the tests were conducted in a dry environment, the principle would work equally well in water. The sensor devices could be installed on an internal robot or carrier pig as an array, and positioned at each of the four quadrants. A practical limitation may be where the joint gap is corroded and/or restricted with encrustation. The use of the technique in this application can be summarised as follows:

- The main benefit is that the technique is simple and effective for monitoring joint gap, and can easily be installed on current internal tools
- The main limitation is that it would require launch and receive facilities and that as-laid data may not be available for comparison
- In conclusion, the technique has potential for monitoring changes in joint gap, but would require further development

IN-PIPE ULTRASONIC INSPECTION OF JOINTS

In-pipe ultrasonic inspection systems (commonly known as in-pipe pigs) used in the oil and gas industries employ a combination of sensors that direct ultrasound pulses radially into the pipe wall and also at an angle to the pipe wall to detect metal loss and cracks in steel pipes. A test programme to investigate the potential performance of ultrasound for the condition assessment of cast iron bell and spigot joints in water mains was carried out by Mecon Limited (Cambridge, UK, www.mecon.ltd.uk), using facilities at Imperial College, London, UK. Data processing and analysis was performed by issquared Limited. The test programme included assessment of the potential for:

- Assessing presence and condition of seal materials
- Joint detection
- Joint gap measurement (pull out and alignment)

Comments are also made on the potential of ultrasonic systems for:

- Detection of bell or spigot corrosion
- Detection of bell or spigot cracking

Test work was performed on a cast iron lead yarn bell and spigot joint. All tests involved scanning an ultrasound probe over the inner surface of the joint with the pipe section filled with water. A number of different frequencies, beam geometries and propagation modes were investigated. Attention was focussed on detecting whether any indication could be found of transmission of the signal from the iron spigot to the lead or yarn behind it.

Ultrasound Scan Setup

The sample joint was a 6 inch (150mm) diameter cast iron lead yarn bell and spigot joint about 12 inches (300mm) long. The inner surface was scrubbed with a wire brush to remove dried corrosion products and organic debris (this allowed performance of the ultrasound technique to be assessed under the most favourable conditions). The end of the bell section of the cast iron joint sample was filed flat and sealed by the use of an adhesive to secure a 10 inch (250mm) diameter plastic disc in place. The sample could then be stood upright and filled with water. An ultrasound probe was then deployed on the end of a scanner arm that allowed it to be rotated about the vertical axis (θ in degrees) and scanned vertically (z direction). Figure 6.26 shows a schematic of the test set up and Figure 6.27 shows the joint sectioned along its axis and includes dimensions to key features.

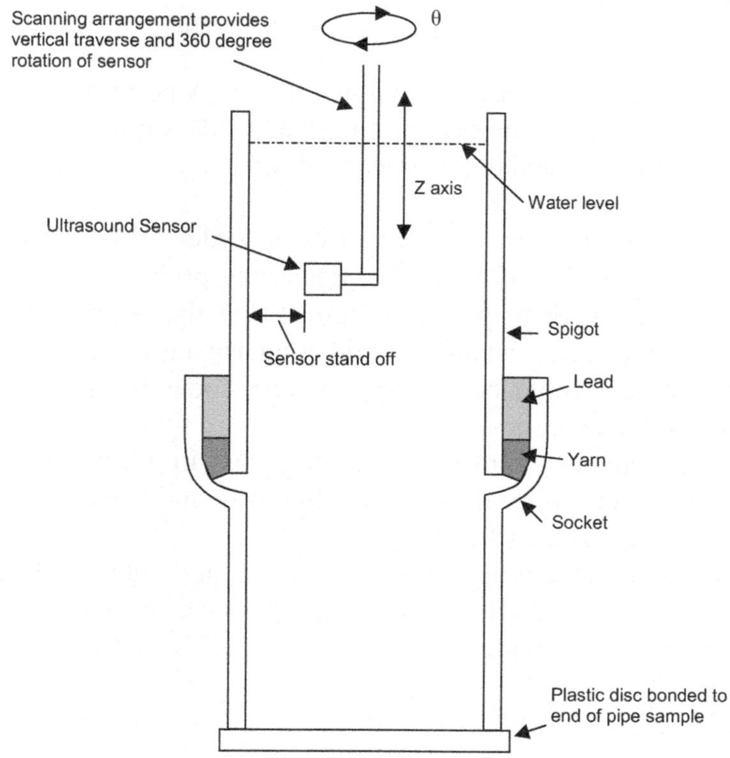

Figure 6.26 Schematic of ultrasound scan set up

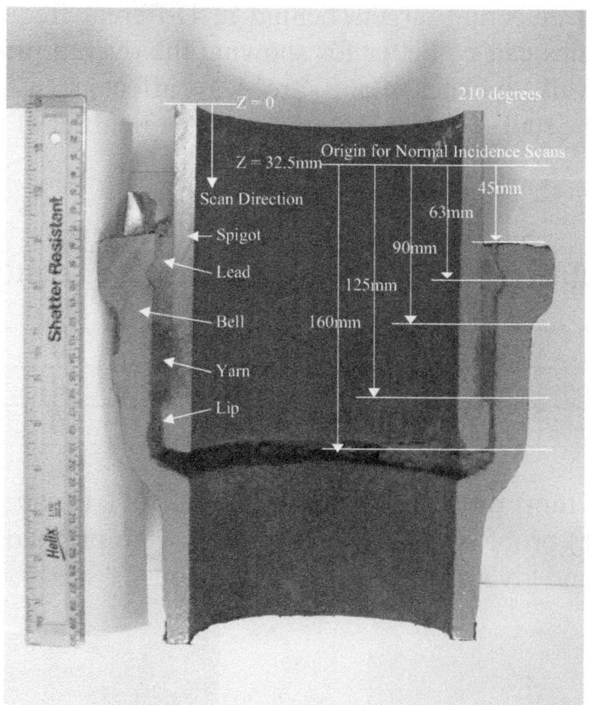

Figure 6.27 Section through cast iron bell and spigot joint

Measurements

The ultrasound scans were carried out as follows. A point on the upper rim of the sample (the top of the spigot end of the sample) was marked as datum, and all scans were referenced to this point. Scans were then carried out as linear z-direction scans separated by regular intervals in θ.

A large number of trial scans were made in order to select the best probes to use at different frequencies and angles of incidence. Focussing probes can sometimes be an advantage because they are less sensitive to probe alignment. On the other hand they will be selective, emphasising targets at the focal distance and de-emphasising targets further away. As it was not clear which focal distance would be of most interest, focussing probes were ruled out and all scans were made using plane probes.

The available frequency range for scanning cast iron pipe is about 0.5MHz to 5MHz. Below this range, it becomes difficult to resolve the front and back faces of the pipe wall. Above it, internal scattering becomes significant.

The sample under test had lightly corroded inner and outer walls, and this further reduced the available range: even at 2.25MHz, scattering from surface roughness was significant. It was therefore decided to limit investigations to the range 0.5MHz to 2.25MHz.

Normal Incidence

The most direct way of gaining information on a metal wall, and any possible targets behind it, is to shine the ultrasound beam directly at it. This gives the simplest return signals and maximises the chances of detecting targets behind it. Different frequencies can yield different information: low frequencies can be better for showing the overall picture (reflector geometries, reflection coefficients, wall thickness) and are less affected by surface roughness. High frequencies can give more detail. Table 6.3 summarises the final set of scans.

Table 6.3
Summary of normal incidence scans

Probe frequency/dia	Stand-off mm	Z start mm	End mm	Z interval mm	Θ start deg	End deg	Θ interval deg
2.25Mhz 0.5in	62	32.5	62.5	2	0	60	2.5
1.0MHz 19mm	37	32.5	62.5	2	0	60	2.5
0.5MHz 19mm	42	32.5	62.5	4	0	60	5

The 0.5 inch (12.7mm) probe was a conventional PZT (Lead Zirconium Titanate, e.g. a stiff piezoelectric ceramic) probe and the 19mm probes were piezocomposite probes. The latter have a higher bandwidth and give better-resolved reflections.

Oblique Incidence

Oblique incidence scans were carried out with the beam incident at 25° to the normal. This is large enough to generate shear waves while avoiding generating compression waves (which would complicate the return signal). The shear waves travel sideways within the pipe wall, and reflections arise wherever propagation conditions change. They will reflect from points at which the properties of the media inside or outside the pipe change. This could include cracks or areas of corrosion. Reflections might thus arise if there is a failure of contact between the spigot wall and the lead fill within the joint, or if there is a film of water between lead and iron.

Longitudinal Oblique Incidence

The probe was angled 25° away from the normal to the wall, pointing downward along the pipe axis. Good signals were obtained using the 1.0MHz probe. Table 6.4 gives a summary of the longitudinal oblique incidence scan.

Table 6.4
Summary of longitudinal oblique incidence scan

Probe frequency/dia	Stand-off mm	Z start mm	End mm	Z interval mm	Θ start deg	end deg	Θ interval deg
1.0MHz 19mm	32	44	74	5	0	60	5

Circumferential Oblique Incidence

The probe was angled circumferentially away from the normal. The 1.0MHz probe could not be fitted into the mounting block at this angle, so the 0.5MHz probe was used. Table 6.5 gives a summary of the scan.

Table 6.5
Summary of the circumferential oblique incidence scan

Probe frequency/dia	Stand-off mm	Z start mm	End mm	Z interval mm	Θ start deg	end deg	Θ interval deg
0.5MHz 19mm	92	40	70	4	0	60	5

Experimental Results

Normal Incidence Scans

Figures 6.28, 6.29 and 6.30 show typical A-scans and B-scans from the 0.5MHz, 1.0MHz and 2.25Mhz scans. An A-scan typically consists of reflections from the inner and outer pipe walls followed by reverberations caused by the transmitted signal bouncing between these two walls. The x and y axes on the A-scan plots represent time of flight and magnitude of reflected signal respectively. A B-scan is made up of A-scans stacked next to each other with a colour scale used to represent magnitudes of the A-scans, the y axis representing time of flight as in the A-scan plots, and the x axis representing axial distance along the joint. In this case, high negative values are represented in blue (reflection nearest the top in the figures), values close to zero in green (bottom of the lines) and high positive values in red (darkest lines throughout figure). The origin (0 mm) for the axial distance on these figures is 32.5 mm from the top of the sample. The higher resolution at higher frequencies is obvious; on the other hand, the general behaviour is clearer at low frequency. The longer, ringier signature of the 2.25MHz PZT probe is evident. All figures show the reverberation within the pipe wall as a regular sequence of signals decaying in amplitude down the page. The signal disappears at the end of the spigot after 165mm in Figure

6.28 before reappearing below the bell. The later reflection at about 25μs is a corner return from the angle between spigot and bell. Figures 6.29 and 6.30 show evidence of a thickening at the end of the spigot: the outer wall reflection is later at an axial distance of 140mm. This corresponds to the lip as shown in Figure 6.27.

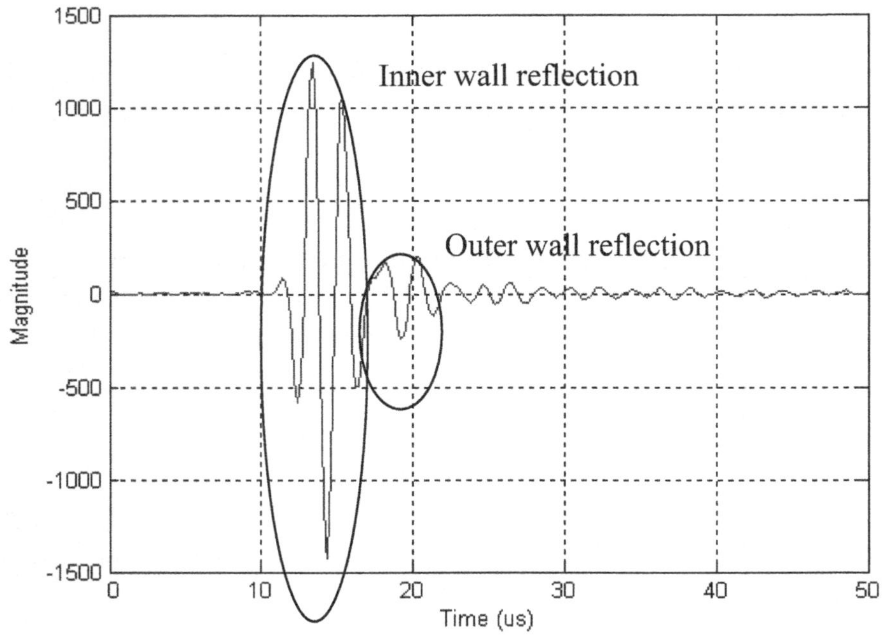

(a) A-scan data

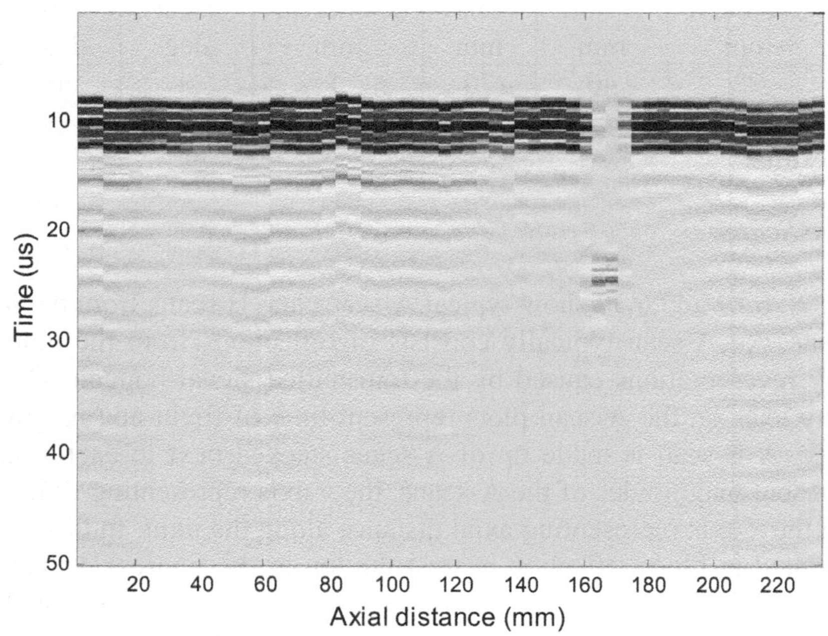

(b) B-scan data

Figure 6.28 Normal incidence 0.5MHz data

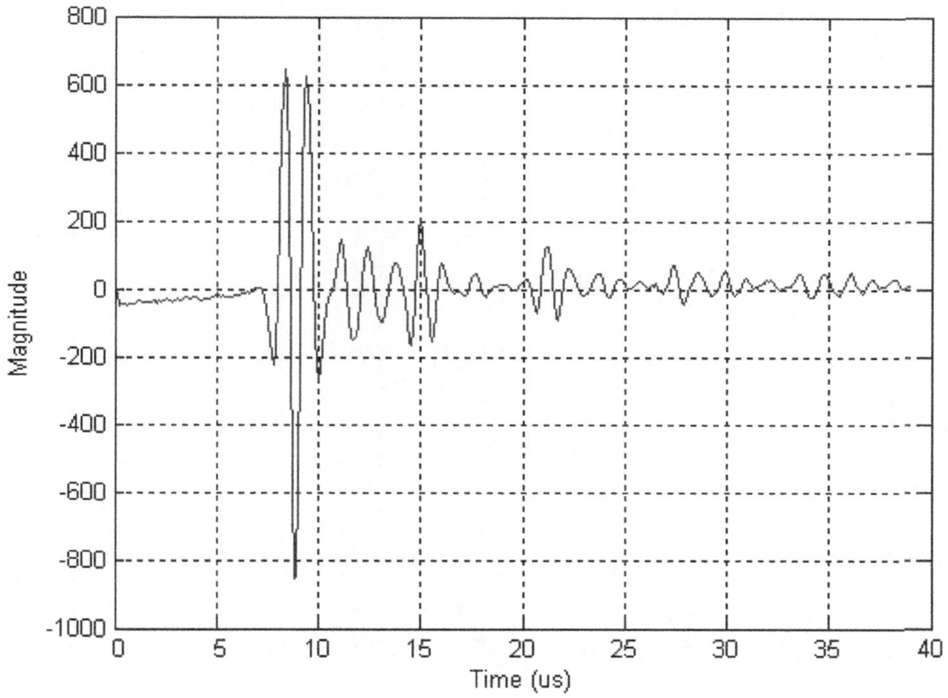

(a) A-scan data

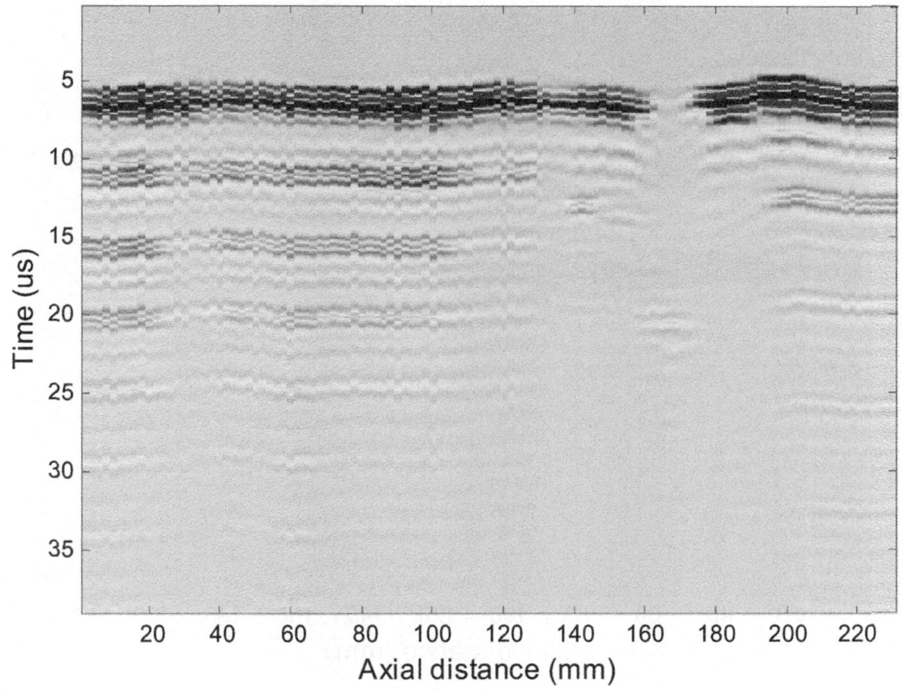

(b) B-scan data

Figure 6.29 Normal incidence 1MHz data

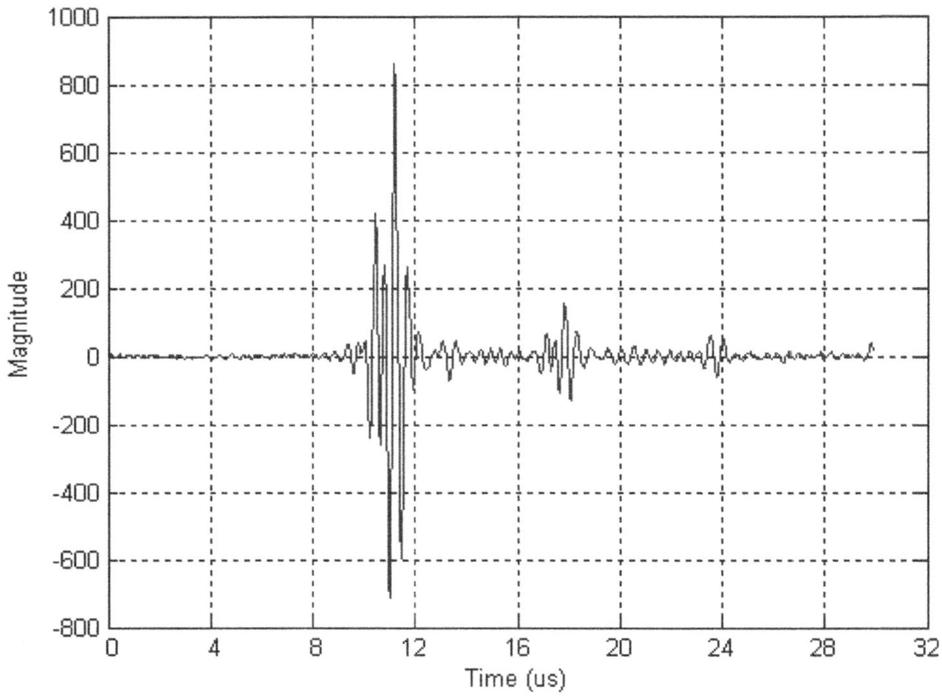

(a) A-scan data

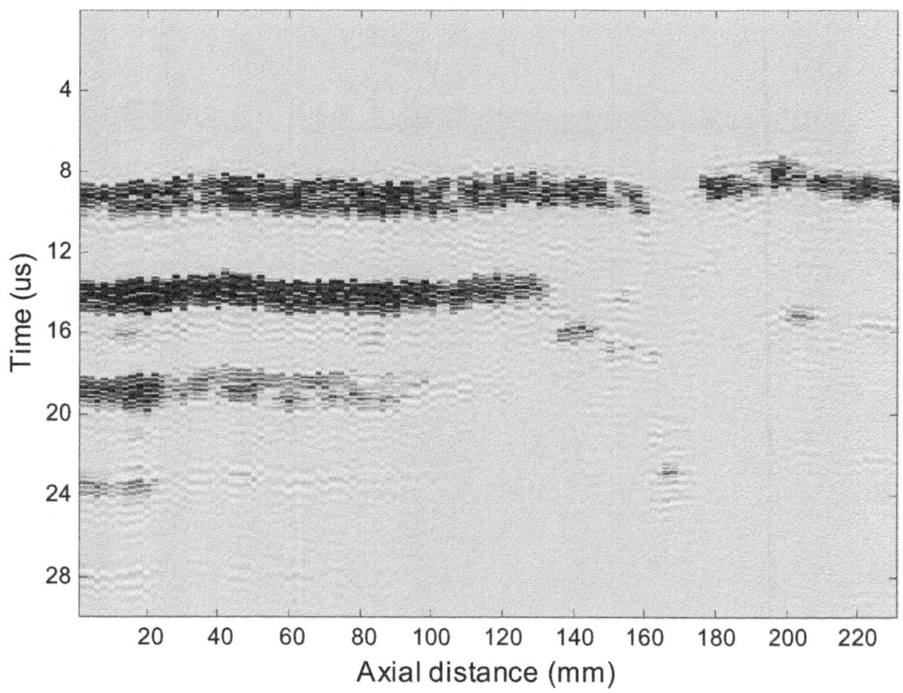

(b) B-scan data

Figure 6.30 Normal incidence 2.25MHz data

The normal incidence data was processed in order to identify any interfaces, i.e. the points at which a change in material properties is encountered (e.g. density of cast iron 7.22 g/cm^3, density of lead 11.2 g/cm^3). At these points, energy is reflected back and identification of this reflection gives the position of the interface. Interface detection was carried out by cross-correlating each of the A-scans with the characteristic pulse from the sensor. The result is a series of peaks that can also be represented in B-scan format. Figure 6.31 shows the B-scan for the 0.5MHz data, at an orientation of 25 degrees. The origin on the horizontal axis is 32.5mm from the top of the sample. At an axial distance of say, 20mm, the first interface at around 10μs is the inner pipe wall, the second interface at about 15μs is the outer wall and the remaining interfaces are reverberations of the outer wall. In the processed data, the increase in wall thickness of the lip at the end of the spigot is very clear at 140mm. The lead would be expected between 40mm and 90mm in Figure 6.31 with the back wall of the lead showing up as additional interfaces among the reverberations; however, none can be seen. In this particular scan the joint gap was filled with debris and appears as a slight lump in the inner wall of the pipe at around 160mm to 165mm.

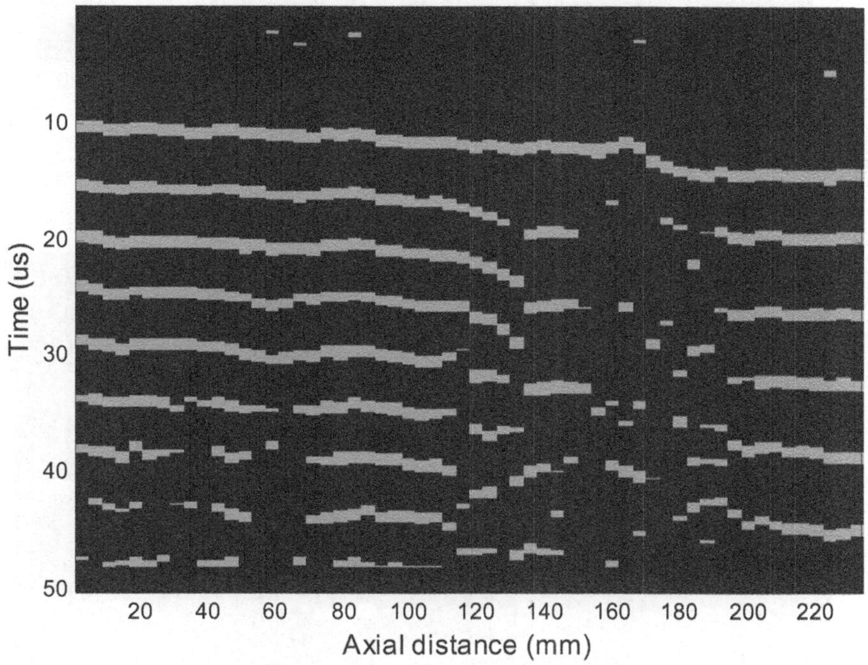

Figure 6.31 Processed data from the 0.5MHz scan

The processed data for the 1.0MHz and 2.25MHz data sets are shown in Figures 6.32 and 6.33 (for the same orientation of 25 degrees). They show similar characteristics as in Figure 6.31, but again, no lead to bell interfaces can be identified.

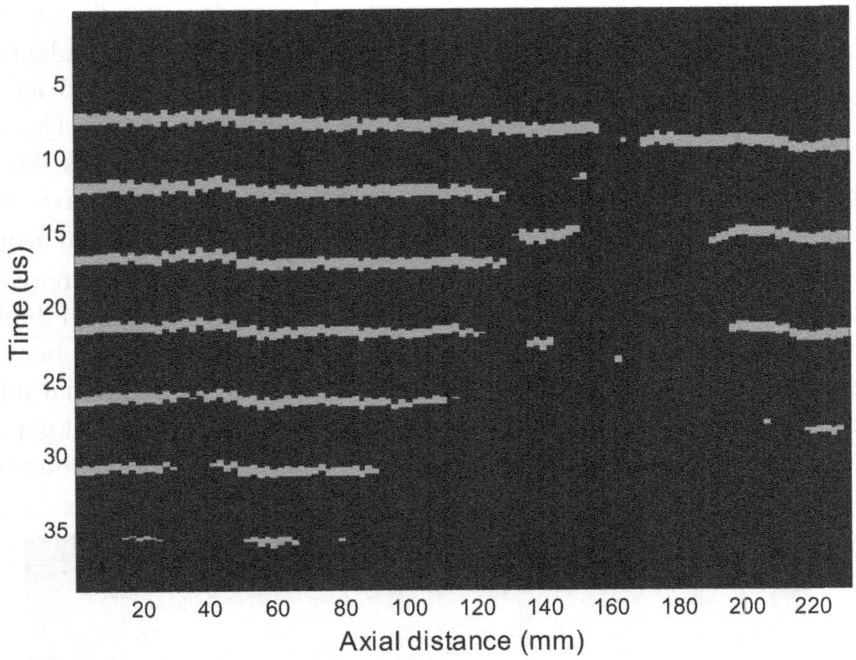

Figure 6.32 Processed data from the 1MHz scan

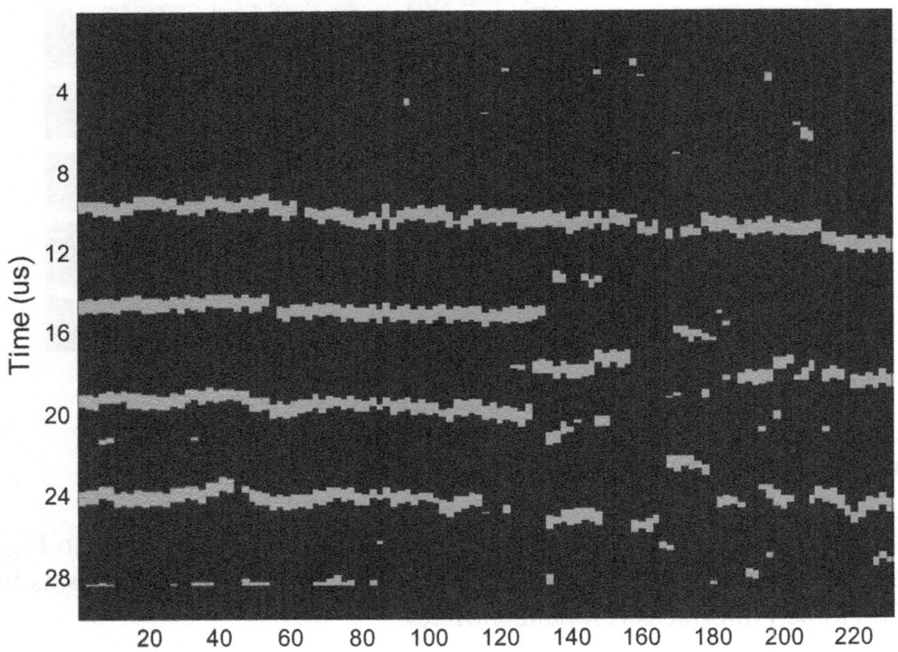

Figure 6.33 Processed data from the 2.25MHz scan

Assessment of all B-scans showed that the semicircular groove in the bell is visible in 16%, 20% and 17% of the 0.5MHz, 1MHz and 2.2MHz scans respectively. A typical B-scan is depicted in Figure 6.34. Evidence of the groove can be seen at an axial distance of 63mm.

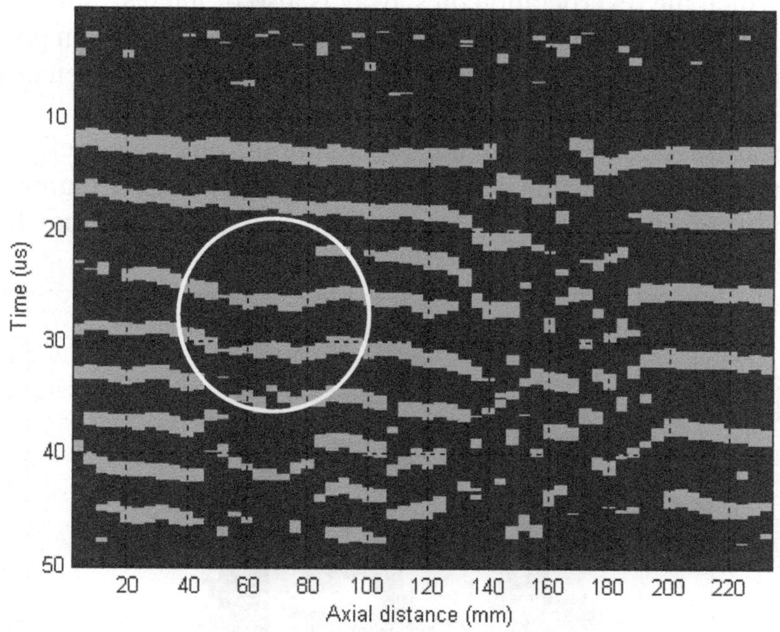

Figure 6.34 B-scan showing semicircular groove in the bell

There is contradictory evidence here. The failure to detect interfaces between reverberations suggests that ultrasound is not penetrating into the lead; however, evidence of the semicircular groove in the bell suggests that it is. An air gap in the groove has also been found in the cross-section of the sample (Figure 6.35); the extent of this gap around the circumference is not known, but it could be the reason why the groove is not seen in all scans. The air in the gap would cause a large proportion of the transmitted signal to be reflected back, but in the absence of the gap, and with the lead and the bell in intimate contact, the reflection due to the change in material density at the interface between the lead and bell may be too weak to be detected. Further tests on a larger number of joints in varying conditions would be required to gain a better understanding of these results.

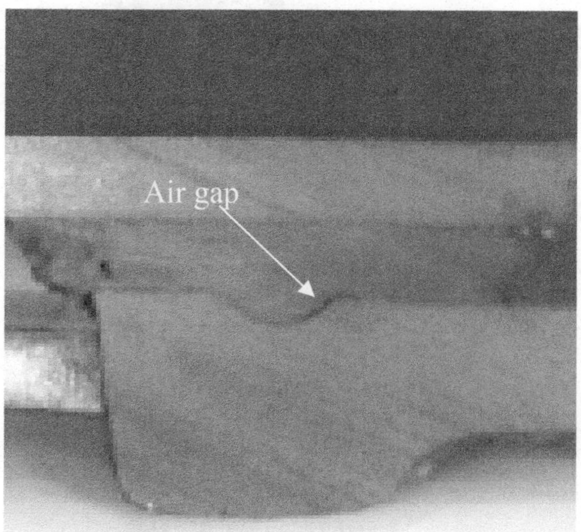

Figure 6.35 Presence of an air gap in the groove

The rate at which the reverberation dies away is also of interest, since it will vary with the reflection coefficients of the front and back wall. Where the back wall is in good physical contact with lead, one would expect a significant reduction in reflection coefficient and hence a more rapidly attenuating reverberation sequence.

A simple measure of this is to map the strength of the outer-wall reflection over the entire surface of the scanned area. Maps for the three frequencies appear in Figures 6.36 to 6.38. The grey scale represents signal amplitude. Once again, the origin is 32.5mm from the top of the sample.

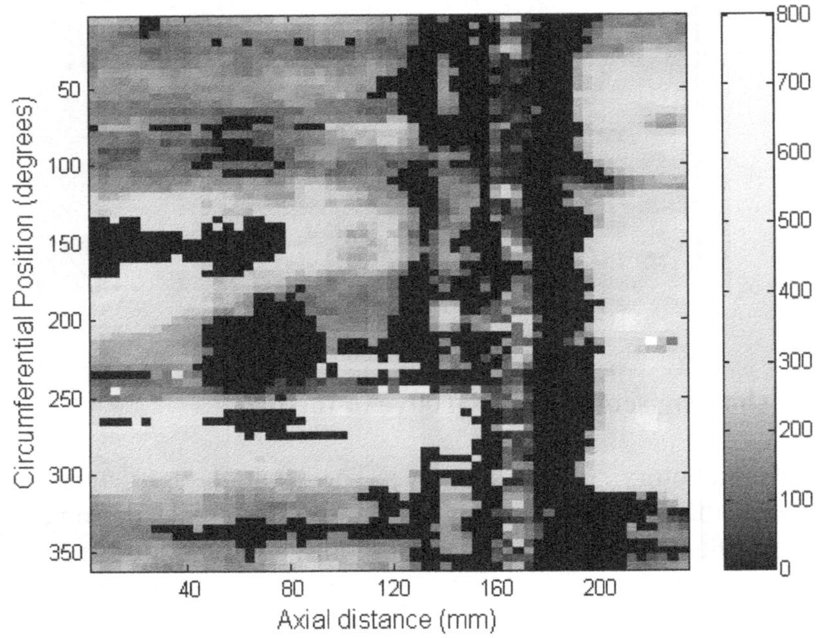

Figure 6.36 Second reflection amplitude map for 0.5MHz data

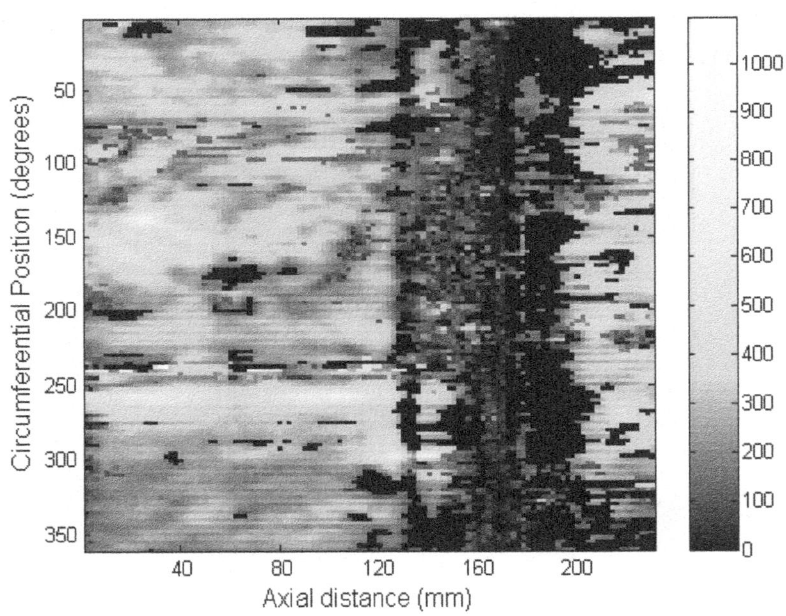

Figure 6.37 Second reflection amplitude map for1MHz data

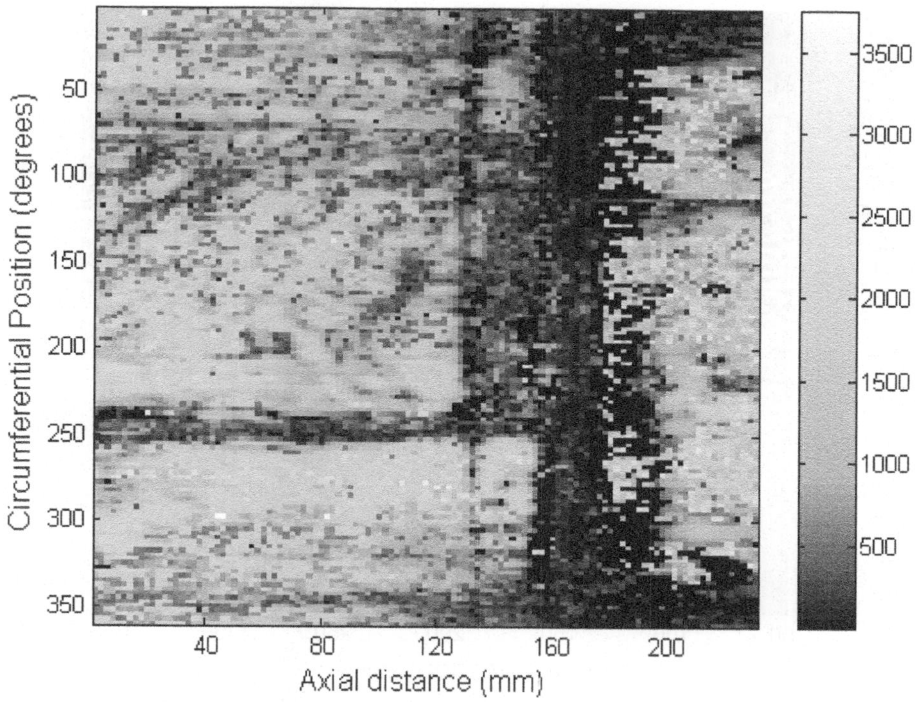

Figure 6.38 Second reflection amplitude map for 2.25MHz data

All three show strong vertical features in the right hand half of the figure, associated with the lip at the end of the spigot, the end of the spigot, and the curve of the bell around the end of the spigot.

It would be expected to see the effects of the lead beyond 45mm. If the lead has a significant effect, a reduction of amplitude would be expected from left to right where lead is present. There is very little evidence of any transition: the only sign can perhaps just be seen on the 0.5MHz scan in Figure 6.36 between 45mm and 90mm, and between circumferential positions of 100 and 330 degrees, but it is far too faint to be relied on. The strong horizontal feature running from the left hand end of the figures at an orientation of 250 degrees corresponded with a slight dip and ridge running axially along the inner wall of the spigot. This appeared to be a casting anomaly. Similarly, there was no indication of the yarn.

Longitudinal Oblique Incidence Scan

The only feature of any interest on these scans was a faint reflection from the outer diameter of the lead fill corresponding to the semi-circular groove in the bell (see Figure 6.27). The B-scan is shown in Figure 6.39. The reflection is visible over most of the circumference beyond 50us.

This feature also appears in the axial B-scans (see Figure 6.40 where the origin on the *x* axis is 44mm from the top of the sample). The two sloping line features between 115mm and 180mm represent the end of the spigot, and the angled surface between the inside diameter of the bell and the inside diameter of the pipe. The sloping line feature at 230mm represents the end of the joint sample.

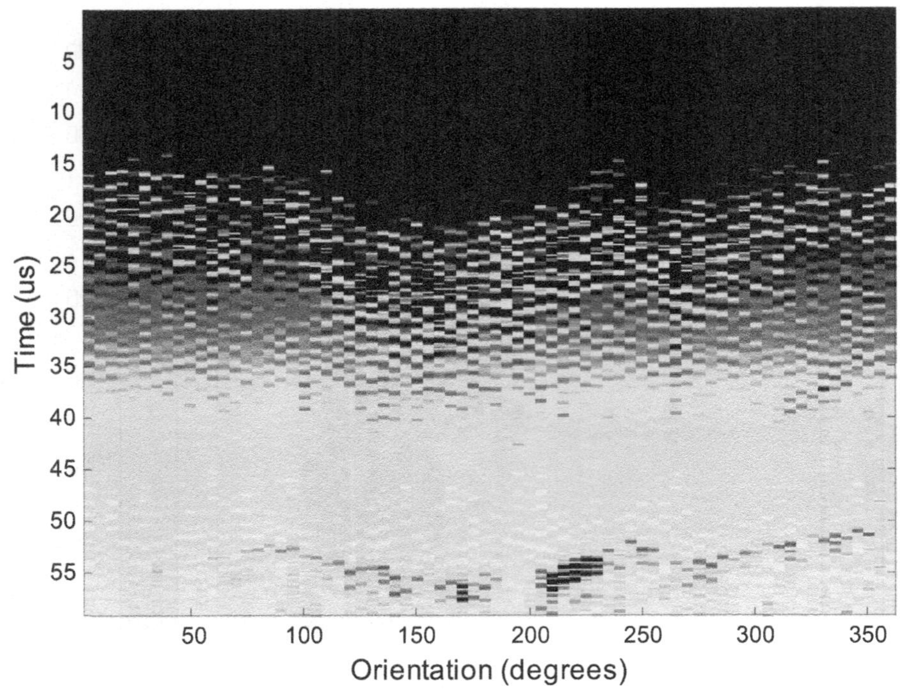

Figure 6.39 Longitudinal oblique incidence data (B-scan around circumference)

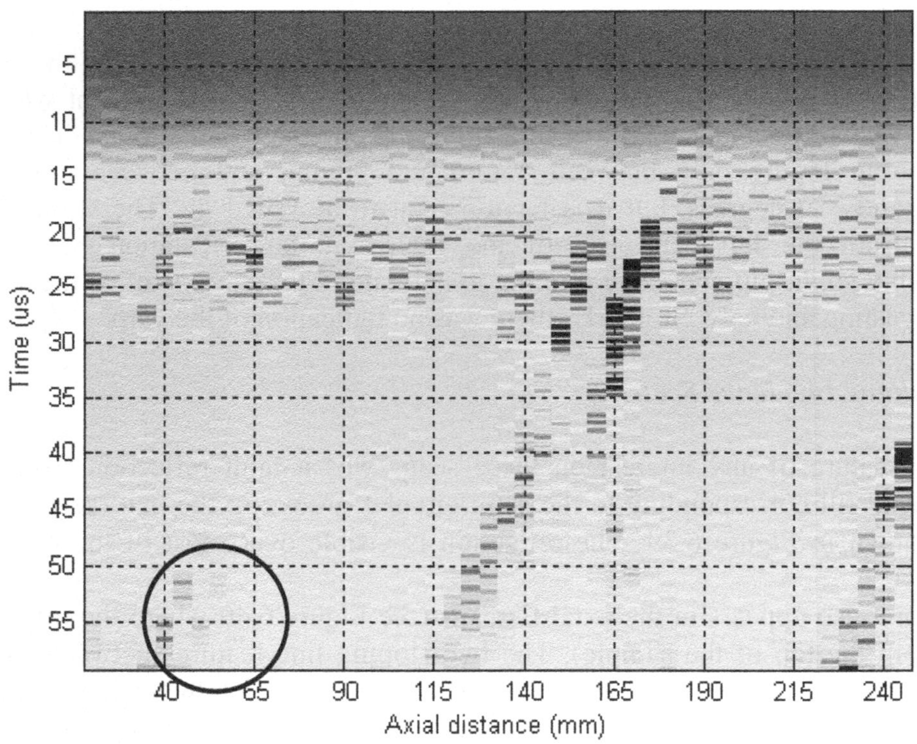

Figure 6.40 Reflection from outer diameter of lead fill at semi-circular groove in bell as seen in an axial B-scan

Circumferential Oblique Incidence Scan

A B-scan around the circumference and 40mm from the top of the sample is shown in Figure 6.41. It depicts reflection events that were typical of the data set as a whole, and represent scattering from surface irregularities in the spigot. There were no features suggesting the presence of the lead or yarn.

Figure 6.42 shows a B-scan along the joint axis with the origin being 40mm from the top of the sample.

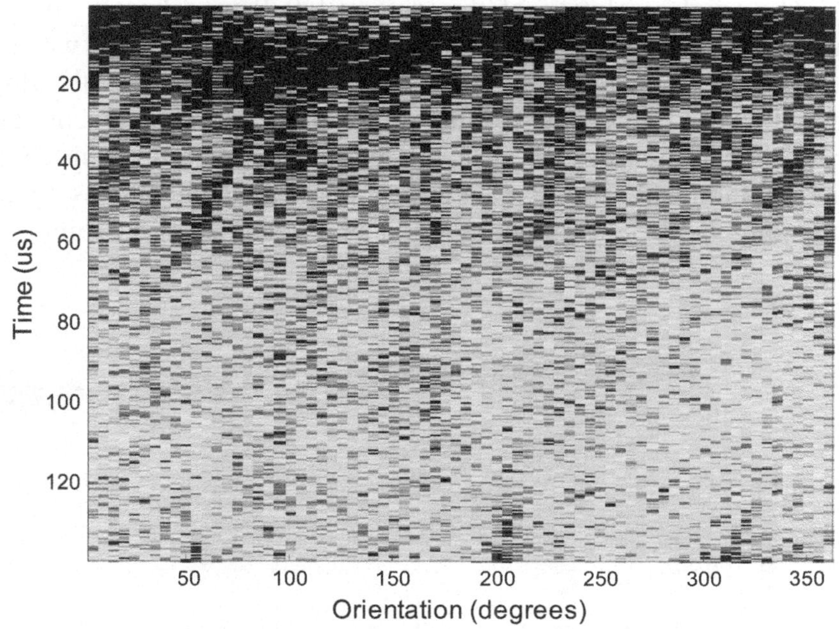

Figure 6.41 Circumferential oblique incidence data (B-scan around circumference)

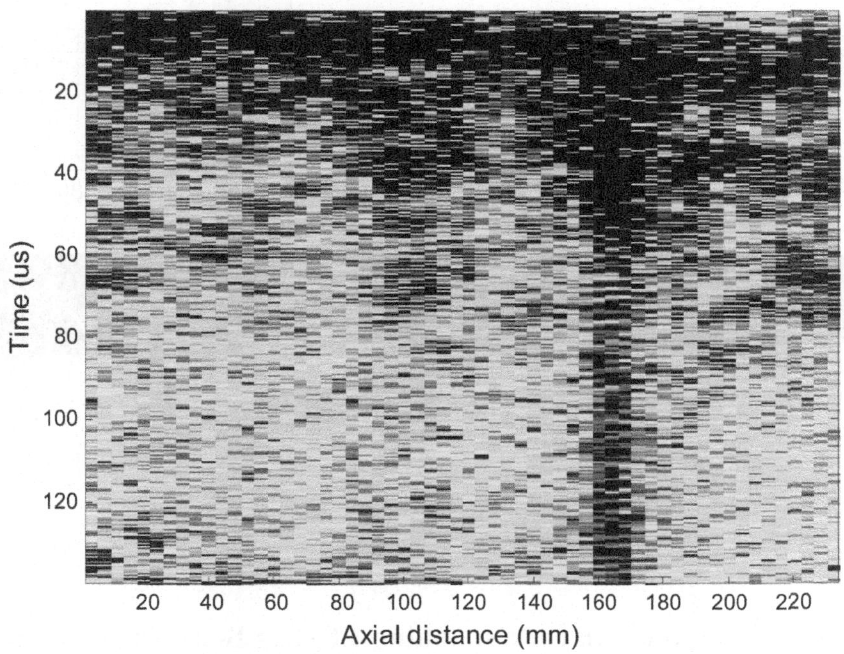

Figure 6.42 Circumferential oblique incidence data (axial B-scan)

The vertical darkish stripe represents the gap at the end of the spigot. There is no evidence higher up of any effect due to the lead or yarn. It was therefore concluded that this scanning mode is not useful for detecting the joint condition.

Joint Detection and Gap Measurement

The joint gap and hence the presence of a joint, can be clearly identified in the normal incidence data. Figure 6.43 shows a section through the joint and corresponding B-scan from the 1.0MHz data. The axial resolution of this data is 0.079 ins (2mm). The missing B-scan data (circled) between 160mm and 176mm represents the joint gap, which was measured at 0.55 ins (14mm) wide. The 176mm taken from the B-scan is around the top of the slope leading up to the corner of the bell in the figure. The resolution of joint gap measurement could be improved by using a finer axial scan pitch for the data collection. This measurement could be used to detect excessive spigot withdrawal or to give an indication of the angle across a joint. In the latter case the width of the joint gap would vary around the circumference of the joint.

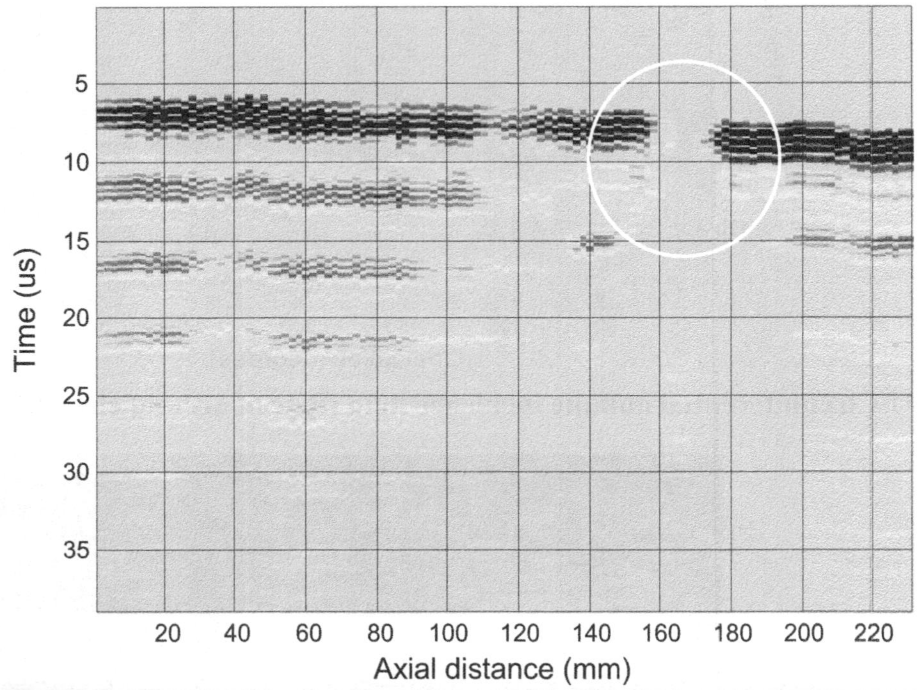

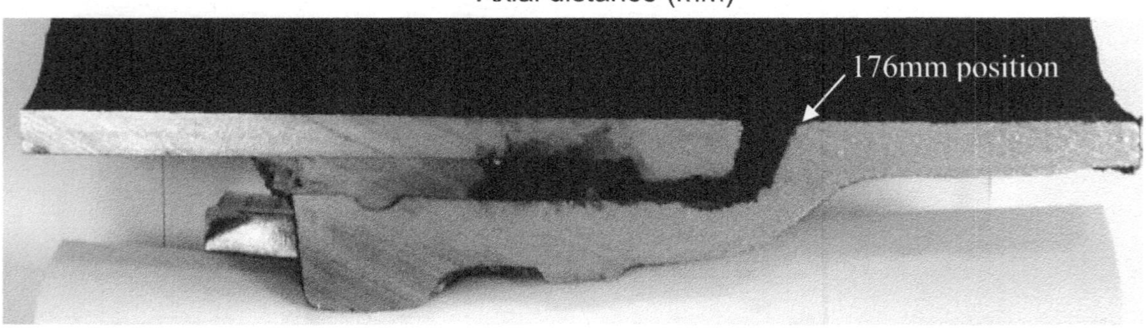

Figure 6.43 Section of joint with corresponding 1.0MHz B-scan showing joint gap

Bell or Spigot Corrosion

Internal or external metal loss would affect the position of the inner and outer wall reflections of the spigot, and hence be visible in the B-scans; effectively giving a measure of wall thickness. This principle is employed by in-line inspection pigs in steel pipes, and ultrasonic probes are also used externally on cast iron water mains (WRc, 1992).

There was no evidence of metal loss in the sample joint. Results from the tests show that it is very unlikely that ultrasound would penetrate through to the outer wall of the bell, and therefore no useful information could be gained on the bell.

Bell or Spigot Cracking

Ultrasonic crack detection pigs are commonly used in the oil and gas industry. This is achieved by directing the ultrasonic probe at an oblique angle to the pipe wall, such that bouncing ultrasonic shear waves propagate along, or around the pipe. Where ultrasonic waves hit an internal or external crack from the side, they are reflected back to the sensor (Wolf, 2001).

There were no cracks evident in the joint sample tested, however, the crack detection principle is illustrated in Figure 6.44 (origin at 44 mm from the top of the sample), which shows a B-scan from the longitudinal oblique incidence test. A crack would be expected to show as a discontinuity in the same way as the end of the spigot shows in Figure 6.44.

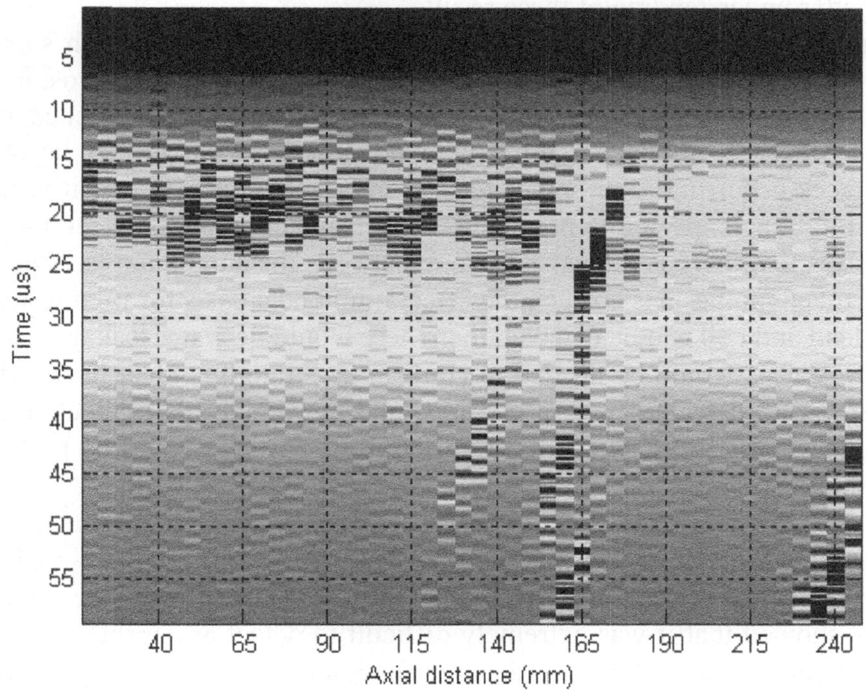

Figure 6.44 Longitudinal oblique incidence B-scan, illustrating discontinuity at a joint

The first sloping line feature (at around 140mm) represents the discontinuity of the end of the spigot producing a reflection of the bouncing shear waves. Although this technique is used to detect cracks in steel pipelines, further testing of joints containing cracks would be necessary to confirm that cracks could be detected in cast iron. The second sloping line feature is caused by direct reflection from the bell, in the vicinity of the joint gap.

Results from the tests show that it is very unlikely that ultrasound would penetrate through to the bell, and it is therefore unlikely that useful information could be gained on cracking of the bell.

Discussion on Ultrasonic Joint Scanning

It has proved extremely difficult to detect any influence of the lead on the ultrasound signals, much less the yarn. Little sign was seen on normal incidence scans, beyond a very slight reduction in the amplitude of the outer wall reflection at 0.5MHz, which was far smaller than variations due to out-of-roundness of the wall. However, there was some indication that the back wall of the lead was being seen in some of the longitudinal B-scans, in the vicinity of the semicircular groove in the bell. Bouncing shear waves propagating parallel to the pipe axis also showed a low-amplitude reflection from the semicircular groove in the bell at a few points. On visual inspection of the joint after sectioning, a slight air gap was seen in the vicinity of the semicircular groove. The air in the gap could cause a large proportion of the transmitted signal to be reflected back, but in the absence of the gap, and with the lead and the bell in intimate contact, the reflection due to the change in material density at the interface between the lead and bell may be too weak to be detected. Further tests on a larger number of joints in varying conditions would be required to gain a better understanding of these results.

The results suggest that there is a layer between the iron and lead, which significantly restricts ultrasonic transmission between them. This could be a layer of corrosion product deposited as a result of electrolytic action between iron and lead in the presence of water. However, the fact that there is almost no ultrasound transmission across this layer (i.e. no reflections from interfaces occurring after spigot to lead interface) suggests that it is not water saturated, since water might be expected to produce at least some coupling between the lead and iron. This in turn suggests that there is no water leakage between the spigot and lead and that the joint may be sound. If this hypothesis is correct, the effect might be a basis for joint testing: if the interface between the lead and bell can be detected then there would be water leakage.

Further laboratory tests would be needed to confirm or refute this idea. It would need to be confirmed that there is no detection in sound joints, and conversely that detection becomes possible when the joint is caused to leak (perhaps artificially). Testing should be performed on joints in a range of conditions.

Conclusions on Ultrasonic Joint Scanning

The test results showed that it was extremely difficult to detect any influence of the lead or the yarn on the ultrasound signals. There was a suggestion that there is a layer between the cast iron and lead, which could be restricting ultrasonic transmission between them, and a hypothesis that the water content of this layer could be used as a measure of joint condition. Further laboratory tests would be needed to confirm or refute this idea.

Test results have shown that the technique could be used for detection of bell and spigot joints, and other forms of joint where a joint gap is present. The technique could also be used to measure joint gap, and hence, spigot withdrawal and the angle across a joint. Corrosion detection was not demonstrated, but ultrasonic techniques are commonly used for this purpose. Crack detection was not demonstrated, but the ultrasound technique used by in-line inspection pigs in steel pipes was illustrated. Further testing on joints containing cracks would be required to confirm that the technique works in cast iron.

The technique has some potential for condition assessment of joints, but would require further testing on a large sample of joints, in varying conditions, before conclusions on its effectiveness could be drawn.

DRILLING OF JOINT SEAL MATERIAL

A patent has been applied for by Consumer Gas, Canada for the insertion of a drilling machine, into a gas main for joint repair application. The device travels along the pipe to a joint and drills through the spigot into the joint cavity. A sealing medium is then injected into the cavity down the flutes of the drill thereby effecting a repair and resealing the joint. This system could therefore offer the opportunity to either take a sample of the joint material for analysis or to test the seal in-situ.

An operational version of the repair system is now available for the gas industry, developed by Engineering Services Inc (ESI). Although this device is essentially for joint rehabilitation, the principle could be developed for joint condition assessment. In order to simulate the principle of the in-pipe drilling machine, and to assess live operational problems, a mounting frame with rack and pinion drive was incorporated on the internal joint repair bogey (Figure 6.45). The device could internally traverse the pipe and when a joint is identified either by a sensor or CCTV, it would be positioned to drill through the spigot. The seal would then be exposed and some hardness test could then be carried. In addition a core sample of the seal could be extracted for analysis, Figure 6.46.

Figure 6.45 Simulated in-pipe drilling machine on bogey

Figure 6.46 Drilled spigot and simulated seal hardness test

Although this technique could be developed to work under live conditions, the problem still remains relating to what information can be gained in respect of the seal/joint condition.

CHAPTER 7
CONDITION ASSESSMENT TECHNIQUES FOR JOINTS WITH DETECTABLE LEAKAGE

EXTERNAL SYSTEMS

Thermography

It has been known for a number of years that leaks from buried pipelines can produce local effects on the soil, and in turn the vegetation. A large proportion of water trunk main networks are located in open ground, making them suitable for leakage screening using airborne remote sensing data. Two types of infrared imagery are applicable.

Thermal infrared imagery provides a measure of the thermal infrared energy radiated from the ground surface and is particularly sensitive to ground temperature variations. Processing of the data allows local surface temperature contrasts and the effects of thermal inertia to be detected, providing evidence of anomalous ground moisture.

Visible and near infrared imagery provides a measure of reflected energy from the ground surface and is particularly suitable for detecting local biomass contrast, standing water and patches of bare ground, all of which can be indicative of ground moisture.

Severe moisture indications, normally associated with standing water, are easily detectable in both Near IR and Thermal IR imagery and can sometimes be seen in aerial photography. Less severe moisture is more likely to be detected in thermal IR imagery. Potential moisture indications that can be seen clearly in individual types of imagery and those that are less clear, but can be seen in a number of types of imagery, should be prioritised for investigation. An example of the identification of moisture using remote sensing imagery is shown in Figure 7.1.

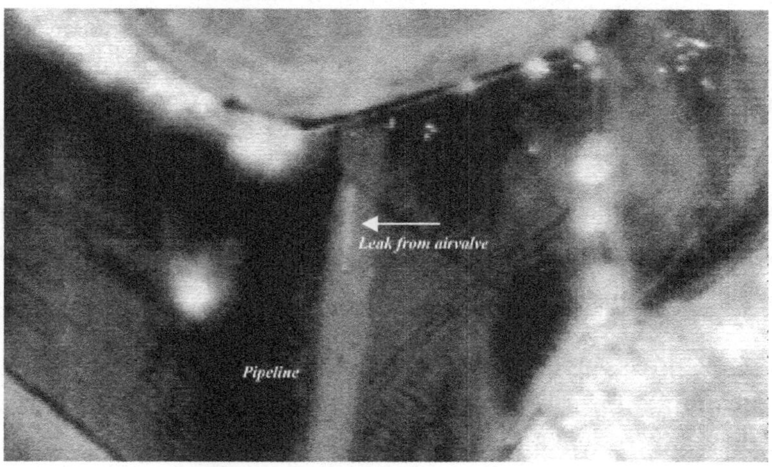

Courtesy issquared

Figure 7.1 Example of aerial infrared scan indicating potential leakage

Any follow-up investigation should first involve discussion with water company engineering staff and landowners familiar with the pipeline, to determine whether there is a known cause for the indication. This should be followed by detailed field investigations where deemed necessary. Such investigations should include visual inspection of the area for obvious causes of the high moisture levels and, where no explanation can be found, the use of leak correlators or in pipe leak survey. It is recommended that site visits be performed during periods of dry weather.

This type of survey can also be performed using thermal instruments suspended above the ground from an appropriate vehicle. It is not considered likely that this method of leak detection would pinpoint the source and location of leaks with sufficient accuracy to determine whether they were from joints or pipe barrel.

Ground Penetrating Radar (GPR)

Theory suggests that the use of GPR to directly detect water pipeline leaks should be possible. The approach would be based on the strong change in soil properties (hence radar propagation properties) produced by saturated soils. Such soils exhibit much increased loss to radar waves, and also a change in velocity of propagation. Thus, the scheme would be to follow the pipeline with the radar, noting the approximate depth and echo strength from the pipe, and looking for a large decrease in echo strength, coupled with a change in the apparent depth of burial of the pipe (shallowing).

The possible problems here involve soil variations not produced by leaks, and/or by real variations in the depth of burial of the pipe. Variations occurring over short distances are perhaps more likely to represent real leaks, but it is the coincidence of the two effects which should be diagnostic.

The technology survey did not identify any companies offering leak detection using GPR as a commercial service. However, it has been reported that GPR has the potential for identifying leaks from buried water mains, either by detecting underground voids created by the leaking water as it circulates near the pipe or by interpreting pipe depth anomalies in GPR images, caused by radar propagation velocity variations due to soil saturation with leaking water (Hunaidi). Experimental work undertaken by Leeds University reported success in the interpretation of leaks from GPR images that identified the leak plume to within tens of centimetres of excavated leak positions (Jackson).

The above studies demonstrated the potential for leak detection by GPR, but neither made definite conclusions about its effectiveness. In particular, as with many GPR applications, soil type and natural moisture content were seen as possible causes of difficulty in interpretation. It appears that development in this area is still ongoing, although no further information is available yet in the public domain. In summary:

- The system shows the potential to detect leaks and pipe joints
- The main limitation is that variations in soil properties and burial depths could give false indications of leaks
- In conclusion, there are no commercial GPR systems available for leak detection

Acoustic Monitoring

The major techniques for detecting water leaking from a pressurised pipeline are based on the noise emitted from the point of leakage. The leaking water emits sound over a range of frequencies. The amplitude and frequency of the sound emitted is related to a number of factors, including:

- Water pressure
- Size of hole
- Pipe material
- Leakage path
- Ground type

Basic location techniques use the ability of the human ear to identify the audible noise within this frequency range. More recent instrument developments have been based on the ability of microcomputer technology to rapidly analyse the noise signals, and to use this data to identify more accurately the position of the leak.

Acoustic Loggers

A recent development in the location of leaks is the use of acoustic data loggers installed in the locality of a suspected leak, Figure 7.2.

Unlike correlators, loggers do not rely on the same noise frequency arriving from two points with a short time delay. They are deployed in numbers over the suspect pipe system and record for several hours the constant source of noise generated by a possible leak. By comparing the sound level and spread recorded at each logger, the user can identify the approximate location of the leak and then focus attention on this area.

Courtesy Northumbrian Water Ltd.

Figure 7.2 Acoustic data logger

The use of this technique in this application can be summarised as follows:

- The main benefit of the system is that it can readily identify an area of leakage, which can then be pinpointed with noise correlators
- The limitations of the system are that it requires experience for interpretation of data, and other techniques will be required to determine if joint or pipe barrel leakage

- In conclusion, the system is widely and successfully used within the water industry to detect leaks

Leak Noise Correlators

Leak noise correlators operate on the principle that characteristic noise emissions from the point of leakage, can be identified using sensors located at some distance away from the point of leakage. In order to identify the location of a leak, sensors are positioned on either side of the suspected leak. The sensors can be either transducers, which are attached to pipe fittings, or hydrophones, which are inserted at tapping points. The noise frequencies from the point of leakage travel along the pipeline at constant velocity, and are therefore received at each sensor after a time delay, which depends upon the distance between the leak and the sensor. The instrument determines the position of the leak, in relation to the two sensors, by comparing the frequency wave patterns received by each sensor, and the time for these to reach each sensor. An example system is shown in Figure 7.3. The equipment operates most successfully on metal pipelines.

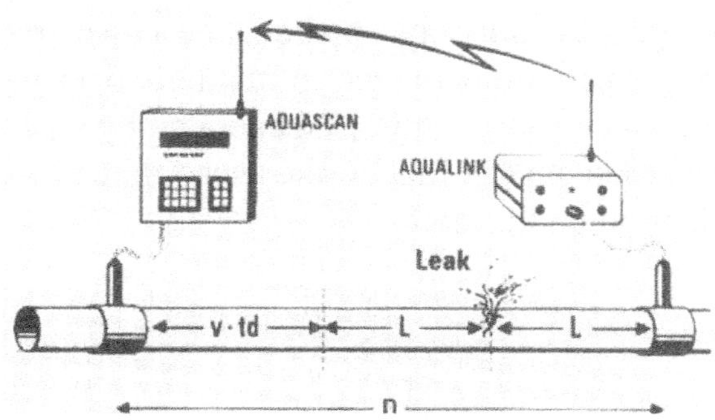

EXAMPLE OF CORRELATOR WITH A RADIO LINK

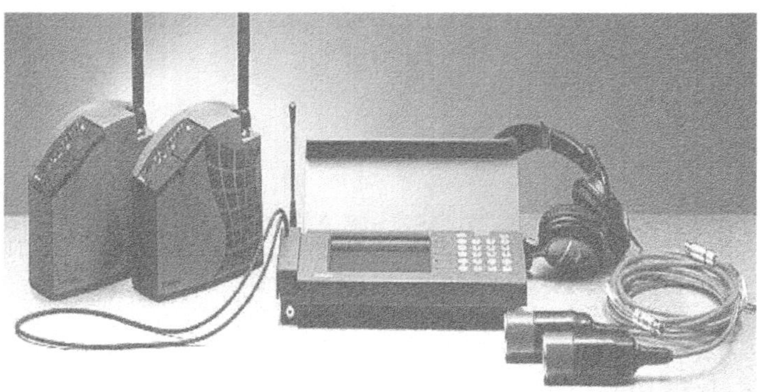

Courtesy Northumbrian Water Ltd.

Figure 7.3 Noise correlator equipment

The use of this technique for this application can be summarised as follows:

- The major benefit is that the system offers leak detection from surface using pipeline features

- The limitations of the system are that other techniques are required to determine if the leak is from the joint or pipe barrel and the system is not effective on large diameter pipes
- In conclusion, the system is widely and successfully used in the water industry on small diameter pipelines

Flow Measurements In-pipe

This method involves the use of a pair of insertion turbine or electromagnetic flow meters, which are inserted at two locations along the pipeline suspected of leaking. Simultaneous measurements are made of the centreline flow velocity at regular intervals over a fixed time period. During this period, the main is operated over a wide flow range if possible. At the end of the fixed time period the meters are interchanged and the procedure repeated for an identical time period. The use of data loggers with each meter enables a large amount of data on simultaneous flows at each meter location to be obtained. A statistical analysis of the recovered data allows flow anomalies caused by leakage, to be identified. The precise location would have to be identified by the use of one of the other techniques discussed in this report. The use of this technique for this application can be summarised as follows:

- The major benefit of the system is that it can readily identify an area of leakage, which can then be pinpointed with noise correlators
- The limitations of the system are that it requires experience for interpretation of data and other techniques are required to determine if joint or pipe barrel leaks
- In conclusion, the system is widely and successfully used in the water industry

Conductivity Surface

This equipment and the technique are fully described in Chapter 5 Joint Detection – Surface, and includes some tests over a wet area with limited success. The following comments relate to possible application for leak detection:

- The major benefit of the technique is that it can identify variations in pipe and ground water conditions.
- The limitation of the system is that it is difficult to interpret results
- In conclusion, the technique has limited potential. No known further research or usage within the water industry

Temperature Probes

This technique uses the possible temperature difference between a volume of subsurface material around the pipeline, which has been contaminated with water from a leak in the pipeline, and a volume of unaffected sub surface material. The temperature probe, would be inserted into the ground where leakage is suspected and the temperature reading noted. Depending upon the season of the year, a rise or decrease in temperature would be noted as the probe is inserted closer to the suspected point of leakage.

This technique would only be suitable for subsurface materials devoid of high concentrations of rock and stones, possibly most applicable where a pipeline has been laid through a rural area under arable land or in sandy conditions. The use of this technique for this application can be summarised as follows:

- No direct benefits have been identified
- The major limitation is that other techniques are required to determine if the leak is from the joint or pipe barrel.
- In conclusion, no known usage

Magic Carpet Acoustic Technique

The Magic Carpet is a leak location device manufactured by Structural Testing Technologies Ltd (S Test) and comprises a 10 ft (3 metre) long durable plastic 'carpet' which is placed over continuous sections of a pipeline. An array of sensors incorporated in the 'carpet' use surface noise correlation to display any leak noise profile. The output is displayed on a LCD screen and an audio output via headphones. This allows the operator to listen in to each sensor response. On hard ground, hemispherical studs are fixed to the sensors whilst in soft ground, spikes are used.

A pipe locator is used initially to locate the line of the pipe. The carpet is then moved along the pipeline in steps, until either a leak is identified or the area of concern has been covered. The equipment is shown in Figure 7.4.

Five rural sites in Northumberland had been identified as possible leak (wet) areas from an aerial leakage-screening search using a thermal infra-red technique, reference AwwaRF project #2612. Northumbrian Water Limited granted permission for a trial of the equipment at these sites. At each site the line of the pipe was identified using GIS maps and pipe locator in the area of the suspected leak. Readings were taken at each position over the suspected distance. No leaks were detected at any of the sites.

There were a number of observations gained from these tests. Firstly the device and associated equipment were cumbersome and heavy to carry to each location and required two people to move it to the next position. Secondly it was very sensitive to background and transient noise, even in the rural environment where a passing car some 330 feet (100 metres) away travelling at a slow speed would saturate the signal. It also took some time to stabilise again following the background noise. This was quite time consuming.

Figure 7.4 Magic Carpet & pipe locator (*Courtesy Structural Testing Technologies*)

A further trial was conducted at Thornaby, Teesside, courtesy of Northumbrian Water Limited, where a leak had been identified by NWL using noise correlators as shown in Figure 7.5. The leak was considered to be in the centre of a 6 ft 6 inch by 1 ft 8 inch (2 metres x 0.5 metres) area.

Courtesy Northumbrian Water Ltd.
Figure 7.5 Leak noise correlator

The Magic Carpet was positioned over the suspected area. In this case, the studs had been attached to the sensors, Figure 7.6.

Courtesy Structural Testing Technologies
Figure 7.6 Magic Carpet and data acquisition equipment

The signal was analysed and indicated that the leak was at the edge of the NWL marked area, adjacent to the manhole. This was observed both visually and audibly. The 'carpet' was moved a short distance and the peak signal again corresponded to the sensor position adjacent to the manhole, Figure 7.7. This confirmed the location.

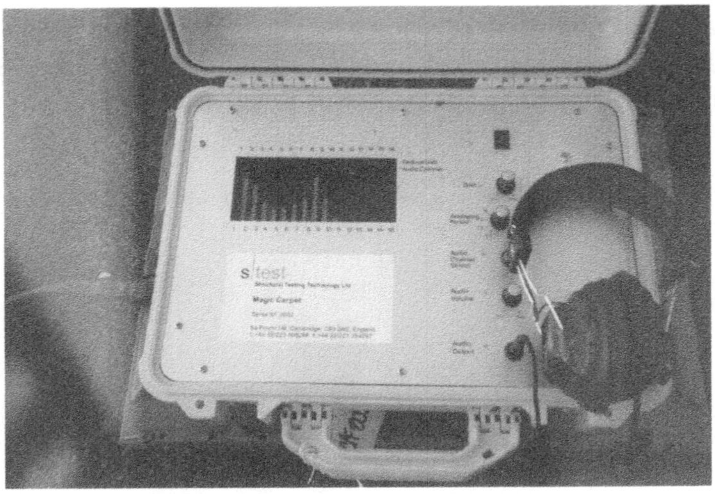

Courtesy Structural Testing Technologies
Figure 7.7 Display showing peak signal.

The 'carpet' was moved along the line of the pipe and about 10 feet (3 metres) away, a second leak signal was identified. However, this was attributed to a branch connection. Subsequent excavation by NWL confirmed that the leak was at the manhole and hence the Magic Carpet had been able to pinpoint the leak. There still was the problem of background/transient noise, even from people moving nearby. Although these tests were conducted on metallic pipes, the system has the advantage of working as effectively on any pipe material exhibiting a leak.

S Test have addressed some of the issues raised from these trials and other tests and have produced a new version of the Magic Carpet based on the same principle. It is shorter at around 5 feet (1.5 metres), has 7 sensors and is much lighter and more manageable. It is claimed that the problem of transient and general background noise has been overcome and the 'carpet' works satisfactorily in an urban environment. However, where there is a constant high-level background noise, > 50 Hz for example from a factory unit, then any leak signal could be swamped. The use of the system in this application can be summarised as follows:

- The main benefit is that the system is claimed to offer an alternative to noise correlators
- The limitations are that it is sensitive to background noise and other techniques are required to distinguish between joint or pipe barrel leakage
- In conclusion, the system does have potential but unknown success rate or usage

INTERNAL SYSTEMS

Sahara Acoustic Technique

The SAHARA system works by inserting a hydrophone through a tapping into the water pipe under-pressure. The hydrophone travels along the pipe attached to an umbilical cable. The signal is passed along the umbilical cable and back out of the main to a signal processing computer. The computer displays the characteristics of the sound source and indicates to the operator when a leak is found. The leak position is indicated on the computer display as a measurement from the insertion point.

The technique is particularly useful for the many distribution and trunk mains where existing methods are unreliable, such as:

- Large diameter mains
- Plastic or other non-metallic mains
- Mains where type of material is uncertain or where there are changes in material
- Mains where there are changes in diameter along the length
- Mains where there are limited connection points
- Mains where multiple leaks are suspected

The device is illustrated in Figure 7.8 with a typical output identifying a leak shown in Figure 7.9.

Courtesy WRc

Figure 7.8 Components of the Sahara system

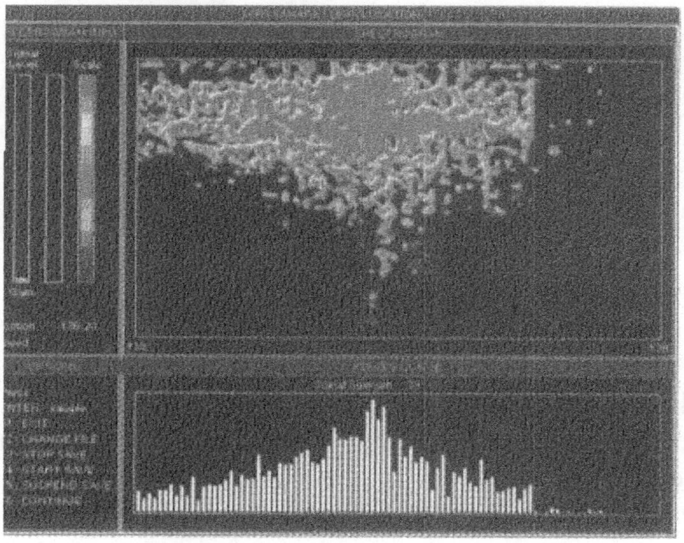

Courtesy WRc

Figure 7.9 Typical output, showing a potential leak

The system has been utilised in a number of pipelines by a number of operators, and a trial was carried out to understand more about the system for use in this application. The Sahara in-pipe leak detection device was used on a DN1200 ductile iron pipe owned by Northumbrian Water Limited at Whittledene Water Treatment works where there was a suspected leak. The pipe runs typically about 6 ft 6 inches (2 metres) depth of cover. The parachute type drogue, which draws in the sensing transducer and trailing umbilical is shown in Figure 7.10.

The pipe was drilled under pressure for a suitable branch connection in order to install the launch tube, Figure 7.11.

Courtesy WRc/Northumbrian Water Ltd.

Figure 7.10 Sahara drogue with sonde and umbilical

Courtesy WRc/Northumbrian Water Ltd

Figure 7.11 Launch installation

Prior to installation and launch, the launch apparatus and drogue and leading umbilical were cleaned in chlorus solution. Thereafter, the umbilical runs through a chlorus (sodium hyperchlorite disinfection solution) bath in the control van, Figure 7.12, which houses the recording and display equipment.

Courtesy WRc
Figure 7.12 Control van with chlorus bath and recording equipment

The device is tracked from the surface by transmitter and receiver transducers, Figure 7.13. The operator follows the line of the received signal and measures distance travelled with a pedometer. The tracking operator and control van are in constant contact so that the position of any features recorded from the leak sonde can be noted, and marked on the surface if necessary.

Around 330 feet (100 metres) of pipe were inspected before the pipe branched at 90°, which the device negotiated without any difficulty. A further 115 feet (35 metres) of pipe were then inspected before reaching the main outlet valve with a different pipe on the outlet. No leak was detected but there was significant noise noted at the valve, which was normal except that it appeared to be greater than expected and of slightly different form. This was attributed to the age of the valve (c1900). It should be noted that subsequent excavation at the main valve revealed a leak equivalent to 10 litres/min.

Courtesy WRc/Northumbrian Water Ltd.
Figure 7.13 Tracking the Sahara device

The system operated satisfactorily on the test run at Whittledene and clearly showed its operational potential. Other operational runs carried out within NWL identified some pipe barrel leaks, which endorsed its potential as a leak detection device. Whether leaks at joints are identifiable or not remains unanswered due to the low noise emitted, but further experience could address this question. The use of the system in this application can be summarised as follows:

- The system can detect small and large leaks in >300mm diameter pipes up to 1.5km from a single launch point
- The limitations of the system are that it is unable to size leaks and it is unable to distinguish between joint and barrel leaks. The system requires access to launch
- In conclusion, the system has been successfully used for leak investigation within the UK water industry rather than for general survey work

Leak Pigs

A leak detection pigging service for liquid pipelines is offered by the BHR Group (www.bhrgroup.co.uk). The service uses an acoustic leak detection pig, based on a development from TUV, Austria (Tscheliesnig). BHR Solutions' HEARLEAK inspection pig is autonomous, containing; power supply, sensors, data acquisition, signal conditioning and data storage. The pig is of a standard cylindrical body and drive/sealing disc design and is driven through the pipe by the flow of liquid. This has the disadvantage that use in water mains may cause detrimental effects on water quality, since the drive discs would disturb loose deposits and tuberculation on the inside wall of unlined metallic pipes. A case study (www.bhrgroup.co.uk/cases/pipehearleak) describing the use of the system in a 27km long HDPE pipe is presented. No leaks were detected during the inspection. In summary:

- The main benefit of the system is that all leaks would be detected over a long pipe run
- The limitations are that launch and receive facilities are required and the pipeline would need to be decommissioned. The system may also scrape loose deposits from the pipe wall, leading to potential water quality problems
- Leak pigs could be utilised in potable water systems

Tracer Gas Injection (all pipe materials)

With this technique, a hydrogen/nitrogen gas mixture is injected into the water main via a standpipe. Since hydrogen gas can penetrate most materials it can be detected at the surface above the pipe with a suitable detector. Radiodetection markets a detection system based on this principle called the 'Sensistor Water Leak Location' system. With this system a surface probe connects electrically to the detector and draws the gas mixture through any surface. For use in soft or wet ground, a ground probe can be used to further enhance detection.

Once the hydrogen tracer gas has been injected into the water pipe, the user follows the route of the pipe with the detector and is alerted to the location of the leak by a sound, which increases in frequency as the leak is approached.

In practice this system has been found to be effective on service lines and water mains up to 12 inch diameter, and can be injected into a live water main or service line. The gas used in this method is a mixture of around 4% hydrogen and 96% nitrogen. This gas mixture will be nonflammable provided the % of hydrogen remains below 5.7%. Potential problems of this technique are considered to include:

- Tracking of gas under road and pavement surfaces, and along outside of pipe.
- Volume of gas required vs pipe diameter vs cost of gas.
- Possible long time delay between injection of gas and permeation to the surface above a leak.

It is not considered that this technique would allow sufficiently accurate pinpointing of leak location to allow discrimination between joint and pipe barrel leakage. It is understood that trials are currently being undertaken using helium as the tracer gas and is showing more success. In summary:

- The system may be relatively low cost for leak detection
- Limited success to date, and could not determine whether joint leakage
- May have potential with helium

CHAPTER 8
POTENTIAL JOINT REHABILITATION TECHNIQUES

There are many potential joint rehabilitation techniques available for potable water, gas distribution and sewer pipes, and many variations on the generic systems. The extent of rehabilitation can vary from replacement to structural rehabilitation through to non-structural rehabilitation and this has been highlighted during the descriptions of the systems below. It may be that the most economic approach is to base rehabilitation technique selection upon the actual condition of the joint, which is determined during the assessment. The generic systems available can be separated into either externally or internally applied repair, rehabilitation or replacement systems.

Externally Applied Repair Systems

The cause of a leak on a cast or ductile iron mechanical joint, is often due to corrosion of the bolts holding the backing ring and sealing system in place. Removal of this sealing force allows the joint to open up and move, thereby increasing the leak over time. In these circumstances, the simplest procedure to reseal the joint, is to replace the corroded bolts, in the recommended sequence, with new bolts. For corroded bolts or joints, consideration could also be given to the application of a retrofit cathodic protection system but this would require access to each joint area and further excavation to install the anode. There are a number of other external repair techniques, which can be applied to leaking, or failed joints and these are summarised below.

Cut-out

In the case of a structural failure (socket or spigot split) or a badly leaking joint, the most usual repair method is to interrupt the supply and cut out the joint (except perhaps for AC pipe where any cutting of the pipe should be maintained to a minimum, and often a full pipe length is replaced to avoid cutting the pipe). This section is then replaced using a short piece of pipe of the same diameter (but not necessarily the same material) with two mechanical repair couplings as shown in Figure 8.1.

Courtesy Northumbrian Water Ltd.

Figure 8.1 Cut-out repair

This method obviously requires extensive excavation around the whole area, but can be used on all types of pipe material up to very large diameter pipes. One of the potential problems with this technique is that couplings of the required diameter have to be available, and for old cast iron pipes, corrosion or the original casting process sometimes makes it difficult to assemble the couplings even if they are the correct nominal diameter. There are also instances where the pipe size may not be standard, and it can take a day or several days to obtain appropriate fittings depending upon the relationship with local suppliers. The maximum size of pipe, which can be repaired in this way, is dependent upon the sizes of the fittings available, and fittings up to 96" (3000 mm) are commonly available.

For PE pipes, there is also the possibility to utilise electrofusion couplings, rather than the traditional mechanical couplings. Electrofusion couplers are essentially a sleeve of polyethylene with a coiled copper heating element just below the internal surface. When a current is passed through the copper heating coil, the polyethylene melts and fusion takes place between the pipe and the coupler. The pipe ends need to be scraped and kept dry and clean. In a repair situation, it is important that the pipe ends are scraped for a distance equal to the length of the coupler. The pipe and fitting have to be clamped to eliminate any movement during the fusion jointing process, as shown in Figure 8.2.

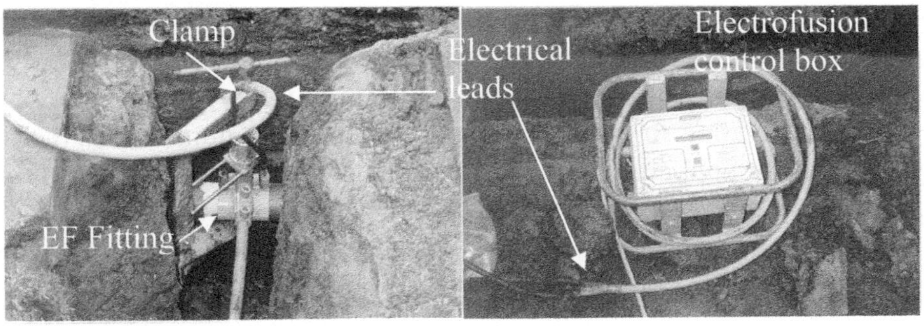

Courtesy Northumbrian Water Ltd

Figure 8.2 Electrofusion jointing and equipment

A purpose made control box usually contains a transformer to supply current to the heating coil embedded in the coupler. A timer automatically switches off the supply after a specified period. After the specified heating time, the coupler must be allowed to cool in the clamps. The couplings are available in diameters of up to around 20 inch (500 mm).

Repair Clamps

Where there is a leak at the pipe joint, a repair clamp of the appropriate size can be used. The clamp completely covers the joint, and has sealing rings incorporated to contain any leakage from the joint. A number of bolts are used to secure the clamp and seals in position as shown in Figure 8.3. There is no interruption to supply unless there is a large leak, where there may be a need to reduce the flow or isolate the section to ensure the clamp is positioned successfully. The system is generally used to repair leaking joint seals, and not structural failures. Again, the clamps are diameter dependent and occasionally, excessive external corrosion product will need to be removed to effect a positive seal with the clamp. Socket repair clamps are generally available for metallic pipelines up to 78 inches (2000 mm).

Courtesy Northumbrian Water Ltd.
Figure 8.3 Typical joint repair clamp

There are several variations of these products, and some clamps incorporate injection points to allow injection of a low viscosity material (e.g. www.furmanite.com - 2003), which helps to seal the leaking joint. A further variation is to apply wrap around shrink sleeves to protect the repair fitting from any further corrosion. An example is shown in Figure 8.4.

Courtesy Northumbrian Water Ltd
Figure 8.4 Shrink sleeve

Repair Collars

Repair collars are also used to re-seal leaking joints by reapplying a load to the sealing material already contained within the joint. The fitting consists of an anchor ring, which is secured to the pipe socket, and an end ring attached to it by a series of bolts. When the bolts are tightened the anchor ring acts in conjunction with the end ring to compress a rubber gasket between the socket and spigot, creating a seal. Again, there is no need to interrupt the supply unless there is a large leak. An example is shown in Figure 8.5.

Courtesy Northumbrian Water Ltd
Figure 8.5 Typical joint repair collar

This fitting is commonly used to repair leaking spigot and socket joints, including lead run joints, with the advantage of not needing to isolate or cut the pipeline. The collars can be applied to pipe joints in ferrous materials and asbestos cement joints. These systems are generally available up to 60 inches (1500 mm) diameter, and are often made to order depending upon the specific requirements.

Potential System – External Encapsulation

Encapsulation of leaking pipe joints is already undertaken in the onshore and offshore gas industries as a repair method. As the method suggests, the excavated joint is encased in a polymeric resin, applied through a soft flexible mould (muff) or by hand. The resin then cures to form a seal around the leaking joint (e.g. www.mwpolymers.co.uk for low-pressure applications). The operation is generally performed with the main still pressurised.

There are several variations of the technique involving the use of different resins and also the reinforcement of the resin using impregnated fibre-glass tape to add strength to the repair (e.g. www.epoxyproducts.com, Fermanite/DML composites www.fdalliance.com). Also metal moulds can be positioned prior to injection of the resin into the annular gap (e.g. www.plcsusa.com). The benefit of these systems is that they do not interrupt supply, and that they can be used to repair leaking joints and pipes with complex features or non-standard diameters. Depending upon the resin used, and approvals being granted for potential contact with potable water, it may be possible to adapt these simple, relatively low cost systems for the repair of leaking joints in water pipes. The main potential concern would be adequate preparation of the pipe surface, to allow appropriate adhesion of the encapsulation material, whilst the joint may be leaking. The higher internal pressure of water distribution pipes may also affect the installation process and long-term integrity of the repair. In summary:

- The main benefit of the system is that it can be applied without interruption to supply and some versions can accommodate complex geometries.
- The limitation is that fabricated steel moulds are required for higher operating pressures.
- In conclusion, the system has been used in various forms for low and high pressure gas pipelines and some use on water pipes for pipe barrel repairs.

Potential System – External Injection

A further external joint repair system is utilised by gas distribution companies. Over the years, the relatively dry natural gas has reduced the capabilities of the yarn to continue to seal the joints, leading to leaks. The lead/yarn joints in cast iron pipes and the rubber sealed mechanical joints in ductile iron pipes can be re-sealed by drilling into the sealant material directly (Consumer Gas Co. Ltd., 1995), or into the bell of the socket just behind the seal (BG, MW Polymers). A polymeric sealant is then injected into the socket either from one position or from several positions on larger diameter joints, whilst the main is still under pressure. The sealant is forced into the area of the leak by a combination of the wicking properties of the sealant around the joint area and the internal pressure. The sealant then cures in-situ, due to the anaerobic (cures in the absence of oxygen) conditions, and reseals the joint. Various grades of sealant can be utilised depending upon the severity of the leak and the operating pressure of the main. The advantage of this system apart from the relatively low cost of the sealant compared to a mechanical fitting, is that only the top of the joint needs to be excavated to allow drilling of the socket. Apart from the reduced excavation cost, the joint is still held by the ground and so is not disturbed any further which can cause the leak to increase.

In water distribution pipes, the leaks may be due to degradation of the seal and joint movement, rather than drying of the yarn. This particular technique and the sealants used, may not therefore be as effective in this environment. However, it may be possible to consider alternative and less benign materials for use in this application, whilst maintaining the benefits of this relatively low cost process. In summary:

- The main benefits of the technique is that the joint only needs to be partially exposed, limited equipment is needed and only small volumes of sealant are required
- The limitation is that it is generally used for pipelines operating at low pressures and the anaerobic sealants would not be suitable for potable water systems
- In conclusion, simple effective technique and with alternative sealants offers potential for joint sealing

Potential System – Clock Spring

Although not initially designed for leaking joints, the Clock Spring pipe reinforcement system (www.clockspring.com 2002) is an interesting concept. The system comprises of a coil shaped composite sleeve (glass fibre and polyester resin), an adhesive and a high compressive strength filler material. The coil shaped sleeve wraps tightly around the external surface of pipes 4 to 56 inch (100 to 1400 mm) diameter, to reinforce the pipe section. The liquid adhesive bonds the composite to the pipe. The Clock Spring Leak Repair System can be used to repair leaks in irregular shapes e.g. bends, fittings and tees but may not be suitable for complex geometries such as joints unless excessive layers of tape are used to even the joint profile. In these applications, a fibreglass tape provides the reinforcement. Some surface preparation may be necessary to maximise adhesion of the reinforcement. This system is similar to those already considered under the external encapsulation section.

Internally Applied Repair Systems

There are many repair and renovation systems, which can be applied internally for both potable water and wastewater pipelines. These can be categorised into local repair systems, pipeline renovation systems and pipeline replacement systems.

Local Repair Systems

These can be described as the repair of local defects on a pipe, without renovating entire lengths of pipeline. Typically they are used for small defective lengths 3 to 9 ft (1 to 3 m).

Joint Sealing. This is a method often undertaken in sewer pipes to prevent infiltration or exfiltration at leaky joints. Typically, an hour glass shaped inflatable ‘packer’ is winched into position so that it is centred on a defective joint, which is located using CCTV. The ‘packer’ is then inflated so that the outer sections of the ‘packer’ seal against the pipe wall leaving an annulus isolated around the joint area. The joint can then be subjected to a pressure test to establish the size of any leak. A grout, gel or resin is then pumped from the centre of the 'packer' into the annulus and the 'packer' remains in place until the sealing material has cured. The 'packer' is then deflated and removed. The 'packers' are available for sewers up to 118 inches (3 m) in diameter. Epoxy resins or epoxy grouts are typically used to repair cast iron, concrete, PVC and AC pipes. An example is shown in Figure 8.6.

Courtesy Northumbrian Water Ltd

Figure 8.6 Internal 'packer' system

The systems are used when the main is isolated and to be economically viable, several joints would need to be treated from one access point. The choice of sealing material will also require careful consideration to meet the necessary approvals. In summary:

- The major benefit is that the technique could be used to treat several joints in non-man entry mains
- The limitation is that the main requires decommissioning and excessive tuberculation would need to be removed
- In conclusion, the technique is used in waste water systems and offers potential in potable water applications

Potential Systems - Other Pig Applied Sealant Systems. There are a number of other similar systems identified through the patent search. One system (patented by Hakko, 1989), highlights the application of a resin based sealing system using a pig applicator. Gas distribution companies (Tokyo and Osaka gas, 1986, 1989) have patented systems, which utilise a series of pigs to apply a slug of resin to a pipe joint. A further system relates to an assembly of two pigs, one of which has pneumatic sealing rings. These isolate the joint for the application of the sealing medium (Willschuetz, K. D., 2000). A further patent refers to an invention primarily for the internal repair of sewers involving a series of bladders and entrained resin (LMK Enterprises).

A further system (Avon Lippiatt and Hobbs Ltd., application ceased 1999), which has been utilised by gas distribution companies, involves the internal application of an anaerobic sealant by means of a spray head which is positioned at the lead/yarn joint by use of a sonde. The spray head can be launched into the main via a 2-inch tapping, and once in position a low viscosity sealant is sprayed around the whole joint area. Several joints can be treated from one insertion point.

This system would be impractical in its current form for use in pressurised water mains due to the spray application method, and the type of sealant used. It may be possible to adopt a similar process for isolated mains, as several joints can be treated from one access point. However, alternative sealant materials may be required to cure in air and comply with water quality regulations. In summary, although the systems have been utilised mainly for low-pressure gas distribution pipelines, they do offer potential for this application. However, the resins may not be suitable for potable water applications and the systems would need to be evaluated for long term sealing performance at higher operating pressures.

Local Lining

This is a similar technique to lining with cured-in-place pipes (CIPP). After cleaning the section of sewer to be repaired, a short length of resin coated lining tube is wrapped around an inflatable packer and winched into the sewer. Typically the cured in place repairs are 3 to 9 ft long (1 to 3 m). The lining tube is usually made from glass fibre or polyester felt, impregnated with an epoxy or polyester resin. Upon reaching the defective section, the packer is expanded until the lining is pressed tightly against the sewer. The resin is cured either at ambient temperature or by circulating hot water or steam within the packer, which is then deflated and withdrawn. Pipe diameters of between 6 and 30 inches (150 to 750 mm) can be repaired using this method.

Alternatively, to conduct a local lining repair on pipe barrel defects or pipe joints, a flexible repair head is winched into the sewer, and positioned across the length of the sewer to be repaired. The repair head is locked into place by expanding a sleeve at either end of the repair head, and a volume of resin is pumped into the isolated annulus forming a short lining. The repair head is left in place until the resin has completely cured. It may be possible to utilise these systems in isolated water mains after drying and preparation of the joints. Again the necessary approvals for the materials would be required, and an evaluation undertaken of long term sealing performance at the higher operating pressures in potable water mains.

Internal Mechanical Sealing

These systems are available for both non-man entry and man-entry pipelines (e.g. greater than 18 inch diameter), and generally comprise of a rubber seal which can be applied internally to water mains or sewers. The type of rubber used for the seal can be tailored to the application. Ethylene Propylene Diene Monomer (EPDM) is available for potable water applications. For man-entry mains, the seal is a profiled annulus and when it is put in place across the joint, two stainless steel annular strips are fitted against its surface. These steel strips have small gaps in them. An example is shown in Figure 8.7.

A hydraulic jack is applied to the gap to force it open, which increases the diameter of the annulus and presses the rubber against the pipe wall to make a leak tight seal. The tightness of the seal is checked by inflating the seal pneumatically and carrying out a soap test (soapy water is sprayed onto the interface between the pipe wall and the seal).

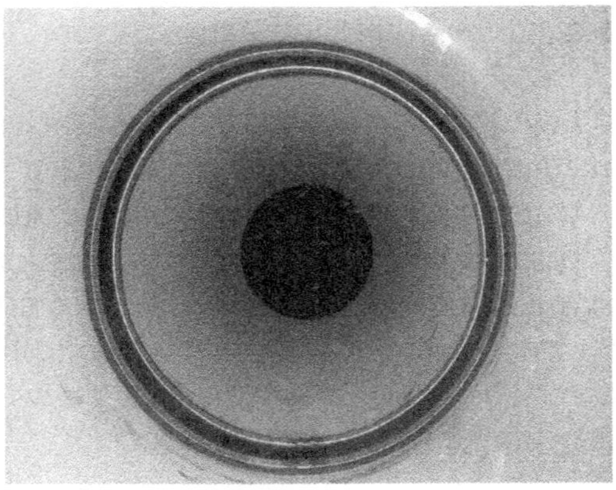

Courtesy Northumbrian Water Ltd

Figure 8.7 Internal mechanical rubber sealing system

The system can also be used to test the integrity of individual joints non-destructively. The seals are generally available in widths of between 4 and 8 inches (100 to 200 mm), but bridging pieces are available with a seal at either end for larger defective areas. Several systems of this generic type are now available.

For non-man entry sewers, there are also mechanical systems available for joint sealing. For example, in the Snap-lock system, the basic repair module consists of rubber seals supported by a cylindrical stainless steel sleeve. The seal is pressed against the inside of the pipe by the circumferential expansion of the sleeve. The seal expands by up to 300% when in contact with water, producing a leak tight seal. There are several variations on the repair module. The first is the basic repair module, which is applied directly to the defective pipe. The second is used where there are protruding roots or encrustation, or where the hydraulic capacity of the sewer is critical. A cutting tool is used to grind a rebate into which the repair module is fitted, making the repair flush with the internal surface of the pipe. A further version consists of a resin impregnated glass fibre sleeve fitted around the outside of the module. The final version consists of a modified module, which allows cementitious or polymeric grout to be injected into voids surrounding the pipe. The system is available in diameters 8 to 24 inches (200 to 600 mm) and can be used in concrete, clay, GRP and PVC pipes. It may be possible to utilise these systems in isolated non-man entry water mains.

A further variation on the mechanical sealing systems is a prototype system under development by AESL. The development has centred around providing a rubber sealing mechanism for defective pipe joints, where the repair sleeve can be positioned whilst the main is still operating normally. A prototype joint location device, insertion unit and repair sleeve has allowed the repair system to be positioned at a leaking joint. The flow of water in the main is used to progress the seal along the main, and deploy the seal at the joint by expanding the collapsed unit, prior to withdrawing the deployment tool back through the insertion unit. The prototype has been operated successfully under water pressure on an 18 inch diameter pipe test loop, but further development would be required to provide a field usable system. A prototype is shown in Figure 8.8.

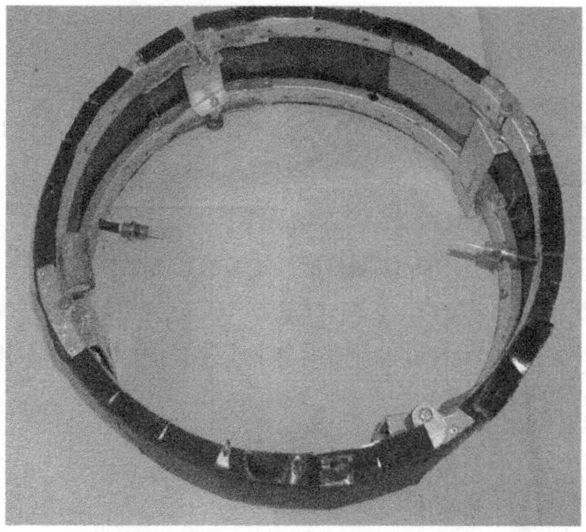

Courtesy Advanced Engineering Solutions Ltd.
Figure 8.8 Prototype remotely applied internal joint sealing system

Other Potential Systems – Mechanical Sealing

Further systems identified through the patent search have highlighted variations on the theme of mechanical sealing. One system (Bridges, D.Y., 1991), relates to a device for internally sealing joints by means of a gasket member tensioned against the sides of the pipe wall by means of adjustable circumferential clamping rings. A similar device also incorporates expansion bands to compress the sealing gasket (Miller Pipeline Corporation, 2000). A further variation relates to a method of sealing by expanding a split sleeve and securing the edges by means of a longitudinal clip device (Johnson Construction Ltd., 1982). A similar device uses an air bag to expand a split sleeve. The sleeve has integral male and female locking strips (Heikki et al., 1990), and (Toa Gurauti Kogyo KK, 1990)

Other mechanical sealing systems incorporate the use of a sealing layer. One patent for a localised pipe repair method uses a series of hinged plate elements. These are contracted to allow ease of insertion into the pipe and then allowed to expand, trapping a sealing layer against the inner wall of the pipe under repair (Toa Gurauti KK, 1996). A further method refers to the use of an internal pipe coil spring, whose diameter has been reduced to allow insertion into the pipe under repair. The spring is then allowed to expand and trap a sealing member against the pipewall and the spring (Owens Corning Fibreglass Corp., 1986). Similar patents have been published using a memory alloy, which expands (NKK Corp., 1990) and a radially expandable composite (Drillflex FR, 1998). In summary:

- The benefits of these systems are that they offer proven long term sealing performance at individual or multiple joints
- Limitations are that the main requires decommissioning and internal access to joints is required
- In conclusion, techniques well known but systems that could operate in non-man entry mains or whilst the main is still in service still need to be fully evaluated

Robotic Repair

To repair defects such as cracks and open joints, remote controlled robots are available to which a number of tools can be attached (www.ka-te-system.com). The robot is lowered into the sewer when under minimum flow conditions and defects are viewed using CCTV. After preparation of the defects, by milling and drilling, resin is applied. After the resin has cured, the repair can be milled flush with the internal surface to optimise flow characteristics. The robots normally operate in clay, concrete or brick pipes, 8 to 32 inches (200 to 800 mm) in diameter. An example is shown in Figure 8.9.

Courtesy Northumbrian Water Ltd

Figure 8.9 Typical robotic type system

A further system, which is currently under development by Innerseal and GTI (Winiewicz and Jarnecke, November 2002), is aimed at perfecting a method for in-situ sealing of cast iron bell joints 16 to 24 inches (400 to 600 mm) in diameter, in gas distribution networks. In-pipe robots will be used along with CCTV to locate the joints, prepare the inner surfaces and seal the joints. Entry to the main is achieved by excavating a hole, approximately 6 feet (1.8 m) wide by 18 feet (5.5 m) long and removing a 4 foot (1.2 m) spool piece from the main followed by capping. A bypass may be necessary if mains are not back fed. An isolation bag is inserted into the main, the end cap is removed, the robot is inserted and a special cap fitting applied. The custom entry fitting facilitates the robot umbilical to travel in and out through a glanding system. A pushing device is used to allow the robot and umbilical to negotiate the main. Once a field useable system has been produced, it is envisaged that further systems could be developed down to 8 to 12 inch (200 to 300 mm) diameter pipe.

Pipeline Renovation Systems

These renovation systems are all designed to install a lining system within the existing defective potable water or wastewater pipeline. Depending upon the system, the liner is non-structural, semi-structural or fully structural that is, it does not require any support from the host pipe against internal and external loading. For most of these systems, any leaking joints are also renovated, but the primary function is to renovate defects within the existing pipe wall. However, if there are a number of leaking joints within the pipeline length under study, it may be economic to utilise a pipeline renovation system rather than repair each individual joint.

Spirally Wound Pipes

In this sewer renovation technique, a ribbed plastic strip is spirally wound by a winding machine, interlocking the strips to form the continuous liner as it is inserted into the defective sewer. The annular space formed between the sewer and the liner may be grouted or the spiral liner can be expanded in-situ to reduce the annular gap to form a close-fit. In this case, no grouting is required. In some systems, the PVC strip is reinforced with a steel band. In larger diameters, the finished liner may be formed on the surface and installed by conventional sliplining methods. Large radius bends and changes in diameter of the existing pipe can be accommodated. The system can be applied to sewer pipes 6 to 118 inches (150 to 3000 mm) in diameter. The completed system does have some structural strength. In summary:

- The benefit of the system is that it allows rapid insertion and provides some structural support
- The limitation is that access to the pipe is required and new connections would need to be made for services
- In conclusion, successfully utilised for wastewater pipe renovation, but may be difficult to apply short lengths in potable water pipelines

Lining with Pipe Segments

In this technique, prefabricated segments are used in man entry wastewater lines to form a new lining within the defective pipe. The short sections are generally used to enable ease of entry via the man holes. The segments are usually sealed at the joints and the annulus grouted to form a bond with the defective pipe. The segmental units are either glass reinforced cement, glass reinforced plastic or plastic reinforced concrete or steel reinforced plastic. In some systems, the units are manufactured in two pieces e.g. the invert and the crown, and are joined together in the sewer before grouting the annulus. The units can be prefabricated for sewers 18 to 196 inches (450 to 5000 mm) in diameter.

Lining with Cured-in-place Pipes

A system in which a thin flexible tube of polyester or glass fibre fabric is impregnated with a thermosetting resin either on-site or in the factory prior to arriving on site. The impregnated liner is then inserted into the defective pipeline either by an inversion or winching process and forced into position against the inner wall of the pipe by water, steam or compressed air. The resin is then cured either by heated water, steam or ultraviolet light. Generally the lining tube consists of a sleeve of porous fabric material that holds a liquid thermosetting resin. A plastic outer layer to the sleeve enables relatively simple and clean handling of the lining tube before and during installation. The inner sleeve can serve a number of purposes, either acting as a carrier for the resin system or also containing reinforcing fibres to enable a composite lining to be installed. A typical system being installed by winch is shown in Figure 8.10. The systems are available for the lining of both potable water and waste water lines of 3 to 60 inches (75 to 1500 mm) in diameter. Similar systems have also been highlighted in patents by Osaka Bodui 1982 and Sumitomo Metal Ind., 1989.

Courtesy Northumbrian Water Ltd

Figure 8.10 Winch-in hoselining system

Reinforced Cementitious Linings

There are two main procedures for installing reinforced cementitious linings into wastewater lines. In the first method, reinforcing steel is fixed to the sewer walls, then steel shuttering sections are locked together, which are centered in the sewer using screw jacks. Concrete is pumped into the space between the sewer wall and the shuttering to form a lining. The shutters are stripped away once the concrete has set and then moved along the sewer to repeat the process. The second process, known as Ferrocement, begins by fixing into position a pre-cast invert panel, which is grouted into place. Steel reinforcing mesh is then fixed to the invert and around the walls of the sewer. The cement mixture is then sprayed onto the mesh in two coats, with a brush finish if required. The systems are designed for man-entry sewer lines and can be used up to very large diameter's.

Spray Applied Linings

This technique can be utilised in both potable water and wastewater lines. The systems apply a lining of either cement mortar or a polymeric resin. The material is applied by either a rotating spray head that is winched through the existing pipeline or by hand spraying directly onto the pipeline's internal surface. Entry is gained to the pipeline at appropriate locations, typically 100m apart. The isolated section of the pipeline is mechanically cleaned and swabbed and then a CCTV survey is conducted to check that the pipeline is free from deposits and that there is no standing water. A high-speed rotating head is winched through the pipeline, centrifugally depositing the lining material over the full internal surface (typically 1mm thick for thermosetting resin and 4 mm thick for cement mortar). Resin can be used for lining potable water lines up to 48 inch (1200 mm) in diameter, and cement mortar in all sizes above 3 inch (75 mm) in diameter. The applied linings are non-structural at this thickness and do not generally seal badly leaking joints.

Thermosetting resin systems are now available which can cure very rapidly. This characteristic may make it possible to overcoat the lining several times to make it more of a structural system, and potentially to be able to seal leaking joints, using several passes of the spraying machine. A typical system is shown in Figure 8.11.

Courtesy Northumbrian Water Ltd

Figure 8.11 Resin lining of cast iron water mains

These systems are generally either non-structural e.g. they require complete support from the host pipe to withstand the internal and external loading, or semi-structural where small holes can be bridged. A system which uses a series of inflatable pigs to distribute epoxy sealant over the inner surface of the pipe is described in a patent by Tokyo gas, 1986 and Frey 1994). In summary:

- The major benefit of these systems is that many joints could be treated from a single entry point
- The limitation is that the main would require decommissioning, the joints may need cleaning or other preparation and materials would need to be applied at a greater thickness
- In conclusion, the system is used in potable water applications but would require further development for treating joints

In-line Pipeline Replacement (or Renovation) Systems

These replacement techniques are used to install a continuous liner within the defective pipeline, when the pipeline has been isolated and dewatered. These systems may be considered if a number of leaking joints are in the same line, or if there are also structural problems with the pipe. They would not be economically viable for individual joint repairs.

Slip lining

This involves insertion into the existing defective pipeline of a single continuous PE pipe lining (Figure 8.12) or jointed sections of GRP (or in some cases DI) pipe to form a continuous pipe within the old pipe. A nose cone is usually fitted to the leading edge to prevent the inserted pipe from "snagging" and to aid attachment of the winch cable. Grouting of the annular space between the existing and new pipe is usually required. The pipes can be jointed either mechanically or by butt fusion for PE. Lining with continuous pipes typically results in a reduced cross section, pipe roughness and usually a net decrease in the pipeline flow capacity. Potable water and waste water pipes 3 to 98 inches (75 to 2500 mm) in diameter can be lined using this technique. Some pre-cleaning may be required prior to insertion

A variation of the technique utilises individual pipe sections, which are shorter than the section to be renovated. The PE, GRP or cementitious pipe sections are lowered into the manhole, pushed into the existing sewer, then further units are added on and jacked, pulled or pushed to the next manhole. Jointing can be push-fit, butt fusion welded or screw threads with 'O' ring seals. Once installed, the annulus between the old and new pipe is filled with cementitious grout.

Courtesy Northumbrian Water Limited

Figure 8.12 Slip lining of cast iron with polyethylene pipe

Lining with Deformed Pipes

This process achieves a close-fit within the pipe by inserting a deformed liner, which is then reshaped within the pipe. The liner is manufactured by firstly extruding the plastic pipe (usually PE, reinforced PE or PVC) and then using a deforming tool to fold the pipe into a compact 'U' shape (Figure 8.13). This reduced cross section ensures easy insertion into the host pipe. After production, the compact liner can be inserted directly into the host pipe or wound onto drums or coils and inserted later. On site, the folded liner will be drawn from the drum and fed into the bore of the old pipe. Once inserted, the deformed pipe is expanded to its original cross-section using a re-rounding device and pressurised water or steam. Pipes from 3 to 60 inches (75 to 1500 mm) in diameter can be installed into defective potable water or wastewater lines.

Courtesy Northumbrian Water Ltd.

Figure 8.13 Lining with deformed PE pipe

Lining with Pipes of Reduced Diameter

This process also achieves a close-fit within a pipe by inserting a mechanically deformed liner, which is then reverted to close to its original dimension within the host pipe. The liner (usually, but not always PE), is reduced in diameter prior to insertion, either by being drawn through a die (Die drawing) or by being rolled in either a hot or cold rolling machine (Rolldown) depending upon the particular process (Figure 8.14). The deformation results in a smaller diameter pipe with either an increased length or an increased wall thickness. The reduction process can take place on site, or the reduced liner can be supplied to site. The liner is then inserted into the host pipe and reverted to its original size, achieving a close-fit. The reversion process can be due to natural recovery of the PE pipe, or by pressurising the liner with water. These deformed PE liners can be successfully installed in old distribution pipes of between 3 and 24 inches (75 to 600 mm) in diameter.

Courtesy of Northumbrian Water Ltd.

Figure 8.14 Rolldown and Die drawing systems

A summary of the approximate diameter range for which the various techniques already discussed are suitable is shown in Table 8.1.

As mentioned earlier in the section, if the condition of the joint could be established, then it may be possible to select a repair or rehabilitation technique, which most closely matches the status of the joint. For example, if it is established that a joint has some deterioration, but is essentially sound, to prevent any further deterioration, it may be possible for example to internally spray apply a resin system only. This action may then be sufficient to maintain the integrity of the joint for the remainder of its service life.

A summary of some of the characteristics of the various joint rehabilitation and repair systems is given in Table 8.2, with an estimation of their potential use depending upon the particular joint condition. Existing systems are classed as those that are currently available for potable water applications, potential systems are already commercially available, but perhaps current use is only in other utility applications and under development systems are classed as those currently being developed for potable water applications.

Table 8.1
Diameter range of potential joint rehabilitation techniques for potable water

Rehabilitation technique	Pipe diameter (inches)						
	3 to 36	48	60	70	96	118	160
Spray applied resin							
Collars							
Clamps							
PE liners							
PE slipliner							
Spray applied cement							
Internal rubber seals							
Resin injection							
Hand applied cement							
Encapsulation							
Couplings							

Table 8.2
Characteristics of the various repair and rehabilitation systems

Technique	Preventative	Leak sealing	Semi-structural	Fully structural
Existing external				
- Couplings				X
- Clamps	X	X	X	X
- Collars	X	X	X	
- Bolt replacement	X	X		
Potential external				
- Resin injection	?	X		
- Encapsulation	X	X	X	
Existing Internal				
- Rubber seals	X	X		
- Spray applied	X	X	X	?
- CIPP	X	X	X	
- Close-fit linings	X	X	X	X
Potential Internal				
- Resin injection	?	X		
- Robot	X	X	X	
- Spiral wound	X	X	X	
Internal under development				
- Live joint repair	X	X		
- Spray applied fibre reinforced	X	X	?	

A **?** indicates that the status is unknown as the system has not been utilised commercially on potable water distribution pipes.

Suggested Parameters to Evaluate Potential Techniques

The various joint repair and rehabilitation options have been fully described earlier in this chapter however, as several of the potential techniques have not yet been extensively utilised within the water industry, it is difficult to evaluate their potential benefits over existing techniques. In order to try and make an assessment, a number of parameters were highlighted and their importance weighted. Each technique was then apportioned a score in conjunction with the weighting, and the summation of these scores provided a final ranking for each technique in comparison with the others. The process is summarised in Table 8.3 with some suggested parameters and scores. These elements can vary considerably and should be tailored to the particular operating requirements of the utility.

The generic repair or rehabilitation techniques to be evaluated were selected either on the basis of their current use within the water industry (e.g. as a type of benchmark for the other techniques to be evaluated against); or for their potential to offer a novel approach to joint repair within the water industry. So for example, one of the current techniques used, internal mechanical sealing using a rubber seal was selected and the various other techniques highlighted in Table 8.3 (none of which are currently used within the water industry for joint repair) were ranked against it. Both externally applied and internally applied systems have been considered.

In Table 8.3, the parameters considered in this example were chosen either on the basis of some of the benefits that the technique can offer to the utility or the ability of the technique to be tailored to the actual joint condition. The parameters are generally in the order in which they may appear in the repair process.

One of the objectives for all utilities is to reduce the number and extent of excavations, not only due to the costs involved, but also due to the disruption they can cause. The weighting for this parameter is therefore high and any reduction in excavation by using the particular technique is therefore offered a higher score. Similarly, if several joints can be treated from one point in the network, this is scored higher.

The reference to whether water quality approval has been granted is highlighted as many of the systems have not been used within the water industry before. The weighting is therefore high, as it is a key concern for some of these systems. Also the scoring is high if the system has already been used in the water industry (perhaps for a different application) or the perception is that the system would not have too many difficulties in meeting the approval requirements (other than the time required to go through the process and the associated cost).

It has been thought necessary to identify the techniques which could be used in non man-entry mains and those which are currently limited to man entry due to the installation technique. That is not to say that these techniques could not be adapted to non man-entry, as it may only be that the current market would not justify the further development costs. For a particular application, the weighting applied for these parameters may need to be increased but for this example, the weighting is lower. What is evident is that there are very few non man-entry repair techniques currently available for use in the water industry and so the potential systems are scored higher if they can be easily applied over a range of joint diameters rather than specific to one diameter.

An operational consideration is how easily the system can be inserted (if necessary) into the water pipe. The higher scoring systems are those which require access to the joint but do not need full bore internal access (which is the most difficult and highest cost to the utility). If the system can be installed without full bore access, this also gives an indication of how readily the

Table 8.3
Example of how a weighting system could be applied to evaluate rehabilitation techniques

Parameter	Weighting	External encapsulation	External resin injection	Internal joint sealing (resin)	CIP liner	Internal seal (mechanical)	AESL tool (mechanical)	Spray applied
Excavation	10	Each joint (1)	Each joint (1)	Several joints (2)	Several joints (2)	Several joints (2)	Several joints (2)	Many joints (3)
Multiple joints treated per excavation/access	10	No (1)	No (1)	Yes (2)	Yes (2)	Yes (2)	Yes (2)	Yes (3)
WQ approval	10	No (1)	No (1)	No (1)	Yes (3)	No (2)	No (2)	Yes (3)
Non-man entry	5	Yes (2)	Yes (2)	Yes (limited 1)	Yes (limited 1)	Yes (2)	Yes (2)	Yes (2)
Man entry	5	Yes (1)	Yes (1)	Yes (1)	Yes (1)	Yes (1)	No (0)	Yes (1)
Ease of access	5	N/A (2)	N/A (2)	Full bore (1)	Full bore (1)	Full bore (1)	Partial bore (2)	Full bore (1)
Main decommissioned	10	No (3)	No (3)	Yes (1)	Yes (1)	Yes (1)	No (3)	Yes (1)
Pre-cleaning required	5	Some external (2)	No (3)	Some internal (2)	Yes (1)	Some internal (2)	No (3)	Yes (1)
Period decommissioned	10	N/A (3)	N/A (3)	Several joints/day (1)	Several joints/day (1)	Many joints/day (2)	Several joints/day (2)	Many joints/day (2)
Preventative	5	Yes (1)	No (0)	Yes (1)	Yes (1)	Yes (1)	Yes (1)	Yes (1)
Leak sealing	5	Yes (1)	Yes (1)	Yes (1)	Yes (1)	Yes (1)	Yes (1)	Yes (1)
Semi structural	5	Yes (1)	No (0)	Yes (1)	No (0)	Yes (1)	Yes (1)	Yes (1)
Fully structural	5	No	No	No	No	No	No	No
System Economics	10	High (3)	High (3)	Low (1)	Low (1)	Medium (2)	Low (1)	Low (1)
Score		170	165	120	130	155	170	180
Ranking		2	3	6	5	4	2	1

system can negotiate any bends or fittings which are already in the pipe. Another operational consideration is whether the main needs to be decommissioned in order to effect the repair. Again this is disruptive and expensive and so warrants a high weighting. Also the period that the main needs to be decommissioned for is a major consideration bearing in mind that these larger diameter mains are usually critical to supply and so the installation of a bypass may be necessary. The systems which do not require the main to be decommissioned, and those which only require the main to be decommissioned for a short period of time, score highest.

A further consideration is the level of cleaning required to enable the repair system to be installed. The cost and logistics of the cleaning equipment, the time required to perform the operation and the disposal of any waste all require consideration within the overall cost of the repair. This can apply to external systems as well as internal systems. The ability of the system to be installed with minimum surface preparation is a major benefit and warrants a higher score.

Other operational considerations can also affect the relative performance lifetime of the repair system and have not been covered in any detail here as they are more applicable to specific individual techniques. However, parameters such as whether the repair is factory or site "manufactured" and site applied can be a big influence on quality and service lifetime. Also whether the system is dependent on good adhesion, or can be influenced by poor weather conditions can also affect long term performance. These issues could be included in any specific evaluation if they are relevant to the specific site conditions.

The remaining parameters all relate to the structural performance of the repair system and how this can be tailored to meet the condition of the joint. For example, if a series of joints are inspected and it is found that the lead/yarn seal is beginning to deteriorate, it may be that if the joints are sprayed with a resinous material to prevent contact with the water (and therefore reduce further degradation), this would be an acceptable solution and prevent the joints leaking in the future. This would be classed as a preventative technique, and avoids the need for excavation and repairs in the future. If the joint already has a small leak, it may be possible to inject a liquid material which, cures in the leak path and re-seals the joint. Again this would remove the need to cut out and replace the joint. This would be classed as leak sealing. The other techniques would apply if the joint was leaking due to damage or degradation of the seal and perhaps movement of the joint. The repair system would need to contain the mains internal pressure. If the repair system requires some support from the joint, then this would be classed as a semi-structural technique, if it can withstand the internal pressure and any subsequent joint movement unsupported, than the system would be classed as fully structural. The ability of the repair or rehabilitation system to fulfil these various requirements relates to the score given. In practice, many factors need to be taken into account. Generally, the cost of a fully structural repair system will be higher than for a preventative repair system but if excavation, decommissioning etc is still required, the overall benefit of using a preventative system is minimal. However, if the system can be easily deployed it could be beneficial to treat the joints either side of an area where a leaking joint has been found and repaired. In this type of activity, the major consideration would be cost per joint, as a large number of joints would be treated within the annual maintenance budget. In other applications, the ability to repair without interruption to supply may be the key requirement or the ability to repair with minimal excavation and disruption to traffic etc.

Within the economics section, there are also many operating parameters to consider such as:

- How many joints can be treated?
- How many operators are required?
- What range of diameters can the system cover?
- How expensive is the equipment required to install the repair system?

- What maintenance costs are involved?

These parameters will vary by application and utility. The economics parameter in this example is really an indication of the simplicity of the system (both the level of equipment required and the ease of effecting a repair). The higher scores are for the simplest joint repair systems.

The overall score is a summation of the individual scores multiplied by each weighting for each parameter. In this example, the internally spray applied resin system has the highest overall score using the parameters and weighting shown in Table 8.3. Fundamentally this is because many joints can be treated from one point and the amount of material applied can be varied to increase the structural strength of the repair. If the major consideration is to treat a large number of joints over a period of time to improve structural integrity then the economics of this system may not be viable and hence the overall score may be lower. These figures have been used as an example of an approach that could be adopted and do not necessarily reflect the weightings and scores a particular utility may use. The parameters chosen and the weightings given therefore need to reflect the major requirements of the particular utility to enable selection of the most appropriate technique.

CHAPTER 9
DECISION SUPPORT TOOL

This section summarises the findings from the technology review and provides a logical approach to selecting tools and techniques for joint detection, inspection, condition assessment and repair. The objective of this section is to:

- Provide a source of reference on potential joint detection, condition assessment and repair techniques
- Provide a logical approach to technique selection
- Assist water companies in obtaining the most economic solutions for a particular repair scheme

The section can be used in its own right, but is also accompanied by a CD containing summary information on the techniques in MS Access 2000 database format, and providing a logical approach to the selection of techniques based on their technical attributes.

JOINT DETECTION TECHNIQUES

This section provides an overview of the technical attributes and economic parameters that should be considered when selecting joint detection techniques.

Selection of Joint Detection Techniques

Figure 9.1 shows the steps to be taken in selecting applicable techniques for any particular joint detection requirement. The terms used are explained in the following sections.

Figure 9.2 lists proven joint detection techniques selected from those covered in this report, together with their key technical attributes. The techniques are not all currently used for joint detection in water pipelines. Some techniques are used in other pipeline industries and some are the subject of research and development. However, all of the techniques are potentially suitable for use in water pipelines. To provide an indication of current status, the joint detection techniques have been allocated status codes, as shown in Figure 9.2. Table 9.1 provides descriptions of the potential techniques covered in this report, together with benefits, limitations, and conclusions on their suitability for use in water systems.

Figure 9.2 can be used to select applicable joint detection techniques for a particular pipeline, based on:

- Pipe material
- Joint type
- Method of deployment of the technique

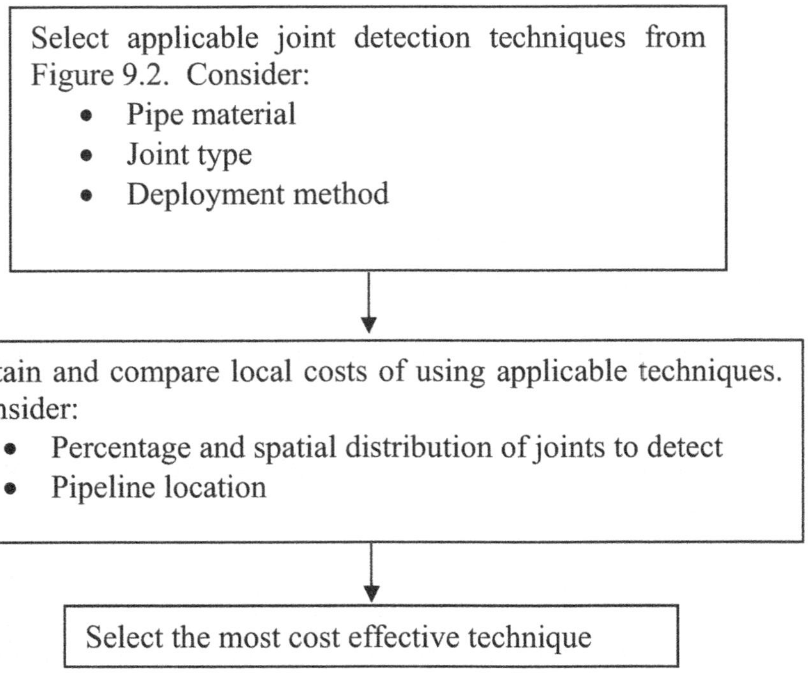

Figure 9.1 Selection of joint detection techniques

Selection based on pipe material and joint type is self-explanatory (note that generic materials and joint types have been used in the table). The methods of deployment to be considered include:

- *Surface* – These techniques can be deployed above ground.
- *On-pipe* – These techniques require intimate contact with the outside of the pipe and therefore a potential requirement for excavation if they are to be used on buried pipes.
- *In-pipe* – These techniques include tethered remotely operated equipment and in-pipe inspection vehicles (commonly known as pigs) that travel along inside the pipe. In all cases, some form of entry and exit to the pipe is required, whether through specially provided entry points or existing points such as hydrants. In-pipe systems would normally require approval for use in potable water pipes.

Deployment restrictions can limit the choice of deployment method. For example, it may not be practical to use surface techniques where access to the pipe is restricted, such as a river crossing or where a pipe runs under buildings.

It is important to consider the reason for detecting the joints, and whether in-pipe detection, or in-pipe detection and surface location are required. For example, the use of in-pipe joint detection techniques is valid if the detection technique is part of an internal repair system (such as detecting the joint and repairing with an internal sealing technique). However, if the objective is to detect joints so that external repair clamps can be fitted, internal detection techniques are only valid if they can provide an above ground location for the joint, so that the joint can subsequently be excavated and repaired. Data from in-pipe pigs fitted with inertial measurement units (often called mapping pigs) can be processed to provide geographic locations of joints. It is also possible to use distance measurements from in-pipe detection systems and to locate joints by measuring corresponding distances above ground.

Factors such as pipe burial depth and ground type should also be considered, since these can affect the performance of some surface deployment joint detection techniques.

In addition to their technical attributes, the cost effectiveness of the applicable techniques should be considered.

Economic Considerations for Selection of Joint Detection Techniques

The cost of deploying joint detection techniques will vary widely around the world, being influenced by local availability, local labour costs, etc. It is therefore necessary to obtain and compare the local costs of using applicable joint detection techniques.

The Percentage and Spatial Distribution of Joints to Detect

The percentage of joints to detect in a section of main and the spatial distribution of these joints can give an indication of whether surface, on-pipe or in-pipe techniques are likely to be most cost effective. This is illustrated by the following examples:

If detecting a small number of joints to see if they correspond to the locations of leaks from a leak survey, surface techniques would normally be the most cost effective. Surface techniques are also the lowest cost to deploy for short sections of buried pipe, since excavation for access to the pipe is not required.

An on-pipe technique may be cost effective for joint detection in short sections of pipe, where above ground access is limited e.g. where pipes run under a building or major road.

If all joints in a long section of pipe are to be detected, then in-pipe techniques may be the most cost effective. This is particularly the case where in-pipe techniques are also being used for joint condition assessment or repair.

Pipeline Location

This has a major influence on the cost of excavation and reinstatement, traffic management and social costs. It is therefore very important to consider the location of pipes (major road, minor road, footpath, verge, open field, etc.) and to use this information when assessing the cost effectiveness of applicable joint detection techniques.

Consideration of the above should allow the cost effectiveness of using applicable joint detection techniques from Figure 9.2 to be compared for any particular local requirement.

		Deployment			Pipe Material			Joint Type			
Joint Detection Technique	Status	Surface	On-pipe	In-pipe	Metallic	Plastic	Mineral based	Push fit	Bolted flange	Welded	PE EFC
Caliper pig	O, P			X	X	X	X	X	X	X	X
CCTV	W, O, S			X	X	X	X	X	X	X	X
Current grading	O, P		X		X			X	X		
Dual frequency radio	P			X	X			X	X		
Guided wave (ferrous)	R, P		X		X			X	X	X	
Hall effect sensors	O, R, P			X	X			X	X	X	
Magnetic locators	P	X			X			X	X		
MFL inspection pig	P			X	X			X	X	X	
Pipe detection and direct measurement	W, O	X			X		X	X	X	X	
Ultrasonic inspection pig	O, P			X	X			X	X		

Status Key	
W	Used in the water industry
O	Used in other industries; gas, oil, petrochemical
S	Used in sewage and waste water pipes
R	Subject of R&D for water industry use
P	Potential system

Figure 9.2 Technical attributes of selected joint detection techniques

Table 9.1
Summary of potential joint detection techniques

Joint detection technique	Description	Benefits	Limitations	Conclusions
Caliper pig (In-pipe)	Caliper sensors measure internal geometry of the pipe and can detect joints and other pipeline features through changes in geometry (e.g. increased bore at joint gap).	All joints located on a long run.	Requires launch and receive facility.	Commonly used in Gas, Oil industry and can detect joints. Unknown usage in Water industry.
CCTV (In-pipe)	Provides images or video of the inside of the pipe. Performance on live water pipes uncertain.	Can identify joints.	Clarity of water/turbulence.	Has potential for use in water industry. Some usage to date in raw water in UK.
Conductivity (Surface)	Hand-held device measuring ground/pipe conductivity, which may detect changes at joints.	Identified pipe and changes in pipe depth and ground conditions.	Awkward to use except in open safe areas. Unable to detect joints.	Unable to detect joints. No known further research.
Current grading system (NYSEG) (On-pipe)	Induced current into pipe leaks to surrounding soils at joints, detectable from surface.	Can detect many joints from one transmission point.	Mixed success due to different ground conditions. Requires excavation or pipe access.	Limited success in US Gas Industry. Some potential in Water industry but requires further development.
Dual frequency radio (In-pipe/Surface)	A dual frequency transmitter is rodded through a pipeline. A low frequency penetrates the pipe wall and can be tracked with a receiver from above ground. A high frequency is also transmitted and this can only be detected when the transmitter passes a push fit or flanged joint.	All joints located on a long run.	Requires launch facility.	Potential technique. No commercial supplier found.
GPR (Surface)	Ground probing radar may be capable of identifying joints where there is an increased diameter at the joint.	Could provide accurate location of joints.	Local variations in soil conditions would affect performance.	No commercially available systems for joint detection. No known research in this area.
Guided Wave – Ferrous (On-pipe)	Transducer bracelet transmits Lamb waves along pipe. Reflected signals are recorded at metallic joints or welds.	Can identify all butt welded joint positions accurately up to 80 metres pipe length in both directions. Non-intrusive.	Single joint only for mechanical joints. Requires excavation for buried pipe.	Limited application for joint detection. No known usage for joint detection only.

(continued)

Table 9.1 (continued)

Joint detection technique	Description	Benefits	Limitations	Conclusions
Guided wave – PE (On-pipe)	Transducer bracelet transmits Lamb waves along pipe. Reflected signals are recorded at plastic joints or welds.	None identified.	Poor transmission of Lamb waves due to signal attenuation in PE. Requires excavation.	No known further research on PE.
Hall Effect Sensors (In-pipe)	Sensors detect changes in flux leakage as they traverse a joint.	Simple and effective. Can be installed on other devices.	Small, corroded and/or debris filled joints can be overlooked or difficult to identify.	Commonly used in UK Gas industry for detection of joints during anaerobic spraying.
Magnetic (Surface)	Hand-held device, which can identify the magnetic orientation of buried pipes and detects end effects and hence joints.	Light and portable. Can identify joints in rural environment with experience.	Interference from other metallic street furniture, vehicles; depth of mains.	Potential for joint detection in rural environment. Usage unknown.
MFL inspection pig (In-pipe)	Well established pipeline inspection tools which detect changes in flux leakage associated with pipeline features (including joints), defects and corrosion.	All joints located on a long run.	Water pipe may require prior cleaning. Requires launch and receive facility.	Commonly used in Gas, Oil industry and can detect joints. Unknown usage in Water industry.
Pipe locator + Direct measurement (Surface)	Hand-held pipe locator and excavation to find first joint, then excavate at spool length intervals.	Operationally finds joints.	Requires skill of pipe locator – success depends upon judgement of length of pipe sections and bends.	Commonly and successfully used in The UK Gas Industry for joint encapsulation work, due to the absence of alternative techniques.
RFEC inspection pig (In-pipe)	This equipment is well established as a technique and can indicate joints and corrosion.	All joints located on a long run.	Water pipe may require prior cleaning. Requires launch and receive facility (some systems can use existing pipe fittings such as hydrants).	Has potential for use in water industry. Some usage in Canadian water industry and Australian Gas industry.
Sonic GRI (Surface)	Hand driven pedometer wheel with Sonar attachment.	None identified.	Unable to cater for different soil conditions.	Project abandoned due to operational limitations. No known further research.
Ultrasonic inspection pig (In-pipe)	Established in-pipe technique for detection of corrosion and cracks. Could be applicable for joint detection.	All joints located on a long run.	Water pipe may require prior cleaning. Requires launch and receive facility.	Experimental work shows that metallic joints with a joint gap could be detected.

JOINT CONDITION ASSESSMENT TECHNIQUES

This section provides an overview of the technical attributes and economic parameters that should be considered when selecting joint condition assessment techniques.

Selection of Joint Condition Assessment Techniques

Figure 9.3 shows the steps to be taken in selecting applicable techniques for any particular joint condition assessment requirement. The terms used are explained in the following sections.

Figure 9.4 lists joint condition assessment techniques covered in this report, together with their key technical attributes. The techniques are not all currently used for joint condition assessment in the water industry. Some techniques are used in other pipeline industries and some are the subject of research and development. However, all of the techniques are potentially suitable for use in water pipelines. To provide an indication of current status, the condition assessment techniques have been allocated status codes, as shown in Figure 9.4. Figure 9.5 shows the condition assessment parameters that can be assessed by each technique. Tables 9.2 and 9.3 provide descriptions of the techniques, together with benefits, limitations and conclusions on their suitability for use in potable water systems. Table 9.2 considers techniques for joints that are not leaking, and Table 9.3 considers techniques for leaking joints.

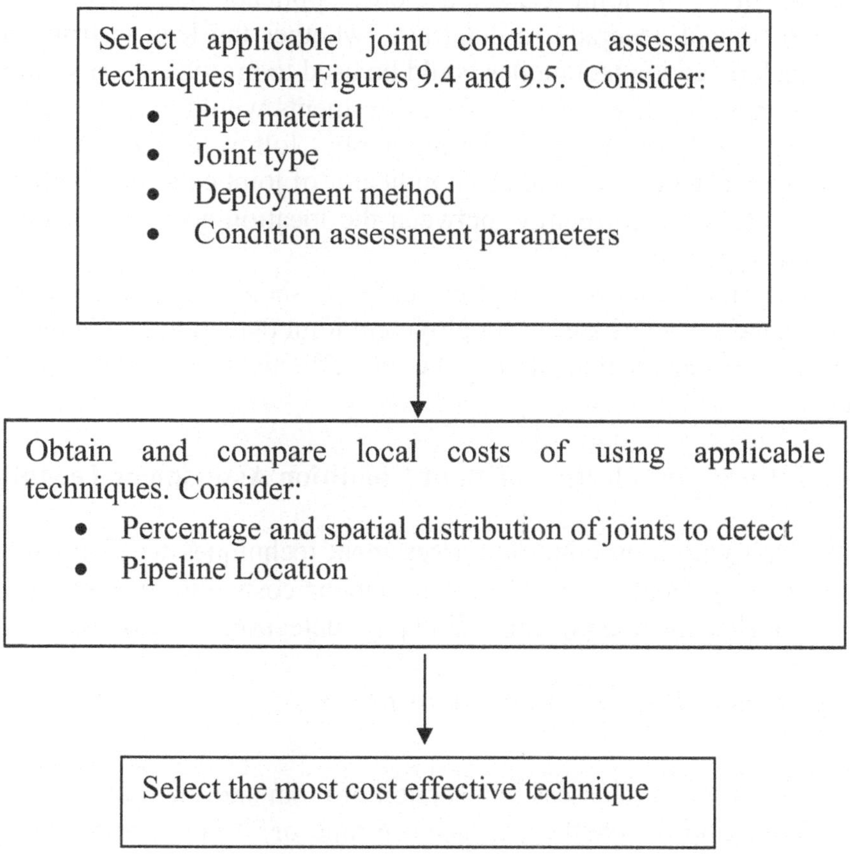

Figure 9.3 Selection of joint condition assessment techniques

Figures 9.4 and 9.5 can be used to select applicable condition assessment techniques for a particular pipeline, based on:

- Pipe material
- Joint type
- Method of deployment of the technique
- Condition assessment parameters

Selection based on pipe material and joint type is self-explanatory, and the methods of deployment to be considered have been discussed above.

Condition Assessment Parameters

Selection based on condition assessment parameters requires some knowledge of the potential failure modes of the joints in a particular area. For example, if joints were failing due to seal deterioration caused by the presence of chloramines in the water, then techniques for assessing seal condition (seal deterioration) would be required. If joints were failing due to changes in angular deflection, then techniques for detecting ground movement could be appropriate.

Techniques for detecting joint leakage are only applicable if the location of the leak can be pinpointed with sufficient accuracy to determine whether the leak is from the joint or pipe barrel. Some of the leak detection techniques could be used to provide a coarse indication of short sections of pipeline containing leaks, but coarse assessments would need to be followed up with accurate leak location techniques (e.g. thermography followed by SAHARA). A further consideration where leak detection is used as an indicator of joint condition is that there must be a high level of confidence in the correlation between the location accuracy of joint detection and leak location techniques.

Factors such as pipe burial depth and ground type should also be considered, since these can affect the performance of some surface deployment joint detection techniques.

In addition to their technical attributes, the cost effectiveness of the applicable techniques should be considered.

Economic Considerations for Selection of Joint Condition Assessment Techniques

The cost of deploying joint condition assessment techniques will vary widely around the world, being influenced by local availability, local labour costs, etc. It is therefore necessary to obtain and compare local costs of using applicable joint detection techniques.

The Percentage and Spatial Distribution of Joints to Assess

The percentage of joints to assess in a section of main and the spatial distribution of these joints can give an indication of whether surface, on-pipe or in-pipe techniques are likely to be most cost effective. This is illustrated by the following examples:

If assessing the condition of a small number of joints to determine whether they are leaking after localised land slip, surface techniques would normally be the most cost effective to deploy, since excavation for access to the pipe is not required.

On-pipe technique may be cost effective for condition assessment of short sections of pipe, where above ground access is limited e.g. where pipes run under a building or major road.

If all joints in a long section of pipe are to be assessed, then in-pipe techniques may be the most cost effective. This is particularly the case where in-pipe techniques are also being used for joint detection.

Pipeline Location

This has a major influence on the cost of excavation and reinstatement, traffic management and social costs. It is therefore very important to consider the location of pipes (major road, minor road, footpath, verge, open field, etc.) and to use this information when assessing the cost effectiveness of applicable joint condition assessment techniques.

Consideration of the above should allow the cost effectiveness of using applicable joint condition assessment techniques from Figure 9.4 to be compared for any particular local requirement.

		Deployment			Pipe Material			Joint Type			
Joint Condition Assessment Technique	Status	Surface	On pipe	In pipe	Metallic	Plastic	Mineral based	Push fit	Bolted flange	Welded	PE EFC
Acoustic loggers	W		X		X	X	X	X	X	X	X
Caliper pig	P			X	X		X	X	X		
CCTV	W, O			X	X	X	X	X	X	X	X
DGPS	P		X		X	X	X	X	X	X	X
Drilling samples	P		X	X	X			X	X		
Flow Measurements	W	X	X		X	X	X	X	X	X	X
GPR	P	X			X	X	X	X	X	X	X
Guided wave	P		X		X						
Hall effect sensors	P			X	X			X	X		
Inertial mapping pig	P			X	X	X	X	X	X	X	X
Laser survey	P	X	X		X	X	X	X	X	X	
Leak noise correlators	W	X	X		X			X	X	X	X
Leak pig	W, O			X	X	X		X	X	X	X
Listening sticks	W		X		X	X	X	X	X	X	X
Magic carpet	W	X			X	X	X	X	X	X	X
MFL inspection pig	O, P			X	X			X	X		
RFEC inspection pig	W, O			X	X			X		X	
SAHARA	W			X	X	X	X	X	X	X	X
Satellite imagery	W, O	X			X	X	X	X	X	X	X
Soil properties	W	X			X		X	X	X	X	
Temperature probes	W	X			X	X	X	X	X	X	X
Thermography	W, O	X			X	X	X	X	X	X	X
Tracer gas injection	W	X			X	X	X	X	X	X	X
Ultrasonic (butt weld)	O, P		X		X	X				X	
Ultrasonic (flanged joint)	P		X		X				X		
Ultrasonic (PE EFC)	O, P		X			X					X
Ultrasonic inspection pig	P			X	X	X		X	X		X
X-ray	P		X	X	X			X	X	X	

Status Key	
W	Used in the water industry
O	Used in other industries; gas, oil, petrochem
S	Used in sewage and waste water pipes
R	Subject of R&D for water industry use
P	Potential system

Figure 9.4 Technical attributes of potential joint condition assessment techniques

	Condition parameter								
Joint Condition Assessment Technique	Joint deflection	Spigot pull out	Crack detection	Corrosion detection	Seal deterioration	Corrosive ground	Ground movement	Leak detection	Weld condition
Acoustic loggers								X	
Caliper pig	X	X							
CCTV	X	X	X						
DGPS	X						X		
Drilling samples					X				
Flow Measurements								X	
GPR								X	
Guided wave				X					X
Hall effect sensors	X	X							
Inertial mapping pig	X								
Laser survey	X						X		
Leak noise correlators								X	
Leak pig								X	
Listening sticks								X	
Magic carpet								X	
MFL inspection pig	X			X					
RFEC inspection pig				X					
SAHARA								X	
Satellite imagery							X		
Soil properties					X	X			
Temperature probes								X	
Thermography								X	
Tracer gas injection								X	
Ultrasonic (butt weld)									X
Ultrasonic (flanged joint)				X	X				
Ultrasonic (PE EFC)				X					X
Ultrasonic inspection pig	X	X	X	X					X
X-ray	X	X	X	X					X

Figure 9.5 Joint condition assessment parameters

Table 9.2
Summary of potential joint condition assessment techniques (non leaking joints)

Joint condition non leaking	Description	Benefits	Limitations	Conclusions
Caliper pig (In-pipe)	Caliper sensors can detect joints and other pipeline features. May have sufficient resolution to measure width of joint gap.	Measurement of joint gap around circumference can provide an indication of joint deflection and spigot withdrawal.	Requires launch and receive facility.	Commonly used in Gas, Oil industry and can detect joints. Unknown usage in Water industry.
CCTV (In-pipe)	Allows visual assessment of internal condition. Possible use of add on tools to assess joint condition.	Visual all round inspection. Possible use of attachments.	Clarity of water/turbulence. Interpretation of results.	Has potential for use in water industry. Some usage to date in raw water in UK.
Differential GPS (Surface)	Established technique for ground movement. May have potential to be applied to exposed pipe for angular displacement.	Periodic monitoring could detect changes in angular displacement at joints.	External access to pipe required.	Potential use as a monitoring system in areas with known ground movement problems.
Guided Wave – Ferrous (On-pipe)	Transmits Lamb waves along pipe. Reflected signals will give indication of joint features.	Can examine several joints from one transmission point on butt welded metallic pipes.	Requires excavation for buried pipes. Not suitable for non butt welded pipes.	Can identify variations in weld performance. Requires further research but has potential.
Hall Sensors (In-pipe)	An array of sensors to quantify angular displacement, misalignment and joint gap.	Simple and effective for monitoring. Simple installation on internal tool.	Requires launch and receive facility.	Has potential for monitoring. Requires development.
Inertial mapping pig (In-pipe)	Established technology for mapping routes of oil & gas pipelines. May be capable of quantifying joint angular displacement and misalignment, depending upon resolution.	Could identify pipe displacement through periodic monitoring of pipe route. Useful in areas susceptible to ground movement.	Would need to run periodically to determine changes in angular displacement.	Potential use as a monitoring system in areas with known ground movement problems.
Laser survey/photograph (Surface)	Application of laser equipment for angular displacement and misalignment.	Simple and effective. Ideal for monitoring above ground pipes and areas of suspected ground movement with buried pipes.	Unknown as laid data. Interpretation of results for joint condition.	Has potential for monitoring areas of ground movement.
MFL inspection pig (In-pipe)	Well established pipeline inspection tools which detect changes in flux leakage associated with pipeline features, defects and corrosion.	Could detect corrosion associated with joints.	Water pipe may require prior cleaning. Requires launch and receive facility.	Commonly used in Gas, Oil industry. Unknown usage in Water industry.

(continued)

Table 9.2 (continued)

Joint condition non leaking	Description	Benefits	Limitations	Conclusions
RFEC inspection pig (In-pipe)	This equipment is well established as an inspection technique.	Claimed joint corrosion detection and pull out.	Water pipe may require prior cleaning. Requires launch and receive facility.	Has potential for use in water industry. Some usage in Canadian water industry and Australian Gas industry.
Robot drilling machine (In-pipe)	Internal pipe robot drills through spigot and injects into joint area.	Could be adapted for assessing Seal hardness, lead, etc.	Interpretation of findings.	Used in US gas industry. Needs development for other potential uses in water industry.
Satellite imagery (Surface)	Well established technique for monitoring ground movement.	Could identify pipes at risk from ground movement.	Assesses risk of joint failure but not joint condition.	Can be used to assess risk of pipe displacement due to ground movement.
Soil properties (Surface)	Established techniques for monitoring soil properties. Monitor water table and other environmental activities.	Simple and cheap to apply for aggressiveness to joint components and gasket material.	Interpretation of results.	Potential use in areas prone to corrosion or gasket degradation. Usage unknown.
Ultrasonic (EF Couplings) (Surface)	Externally applied technique used for QA testing of PE Electrofusion Couplings.	Can detect lack of fusion, inclusions and over heated joints	Requires external access to the pipe.	Currently used on steel pipes and PE gas pipes. No technical reason why technique could not be applied to water pipes.
Ultrasonic (Flanged Joints) (Surface)	Transmitted ultrasonic signal across flange assessed for different gasket conditions and loading.	Simple and effective.	For use on full diameter flat faced flanges only.	Can identify changes in gasket condition but requires further research and development. May have potential in the water industry.
Ultrasonic (PE Butt welds) (Surface)	Externally applied technique used for QA inspection of butt welds.	Proven technique for QA in the gas industry.	Some systems have poor performance in detecting lack of fusion of PE welds. Requires external access to the pipe.	Primarily used on PE gas pipes, No technical reason why it could not be applied to water pipes.

(continued)

Table 9.2 (continued)

Joint condition non leaking	Description	Benefits	Limitations	Conclusion
Ultrasonic inspection pig (In-pipe)	Proven in pipe technology for pipe barrel condition and crack detection.	Has the potential for measurement of joint gap (pull out and alignment), corrosion, cracking and presence of seal materials.	Water pipe may require prior cleaning. Requires launch and receive facility. Development required for use in water pipes.	In pipe systems commonly used in the oil, gas industries. Some experimental work done for this report shows potential of technique for assessment of joints in water pipes.
X-ray – External (Surface)	X-ray capabilities for assessing joint condition.	Can identify cracks, areas of corrosion and partial leak paths.	Operationally difficult to deploy. Requires excavation.	Requires development for this application. Usage unknown.

Table 9.3
Summary of potential joint condition assessment techniques (leaking joints)

Joint condition leaking	Description	Benefits	Limitations	Conclusion
Acoustic monitoring/loggers (On-pipe)	Well established leak location technique, deploying several loggers and monitoring noise levels and frequency differences.	Can readily identify area of leakage. Pinpoint with noise correlators.	Requires experience for interpretation of data. Other techniques required to determine if joint or barrel.	Widely and successfully used in Water industry.
Conductivity assessment (Surface)	Measurement of conductivity would show up changes in value due to the presence of water.	Can identify variations in pipe and ground water conditions.	Difficult to interpret results.	Limited potential. No known further research or usage.
Flow measurements (On-pipe)	This measurement along a pipeline can detect flow differences, which could be due to leakage from some source.	Can readily identify area of leakage. Pinpoint with noise correlators.	Requires experience for interpretation of data. Other techniques required to determine if joint or barrel.	Widely and successfully used in Water industry.
GPR (Surface)	Latest research shows GPR may be able to detect leakage.	Potential to detect leaks and pipe joints.	Variations in soil properties and burial depths could give false indications of leaks.	No commercial systems available for leak detection. Subject of some research for water industry use.

(continued)

Table 9.3 (continued)

Joint condition leaking	Description	Benefits	Limitation	Conclusion
Leak pig (In-pipe)	Inspection pig that detects leaks using acoustic techniques.	Can detect leaks in a long length of pipeline. Suitable for pipe from 4” to 16”.	Pig would disturb tuberculation and pipe wall deposits in metallic pipes. Requires launch and receive facilities.	Leak pig has been used for assessment of new build water pipelines, but is considered too aggressive for existing pipelines.
Listening sticks (Surface/On-pipe)	Hardwood sticks with a shaped earpiece at one end can be used to detect leaks. These can be used in direct contact with the pipe or fittings, or on hard surfaces above a buried pipe. Sound intensity increases with proximity to a leak.	Can be used to pinpoint the position of a leak.	Requires skilled operator. Best performance at night when background noise is at a minimum.	Widely and successfully used in Water industry.
Magic Carpet (Surface)	An array of sensors on the surface detect acoustic vibrations from leaks.	Claimed to be an alternative to noise correlators.	Sensitive to background noise Other techniques required to determine if joint or barrel leak.	Has potential. Unknown success rate or usage.
Noise Correlators (Surface)	Well-established leak location technique based on constant velocity of signal and time intervals.	Simple and effective leak detector from surface using pipeline features.	Other techniques required to determine if joint or barrel leak. Not effective on large diameter pipes.	Widely and successfully used in Water industry.
Sahara (In-pipe)	This is a tethered internal pipe inspection tool, which can, by the use of acoustic techniques, detect leaks.	Can detect small and large leaks in pipes up to 1.5 km. Suitable for pipes > 300mm.	Unable to size leaks. Unable to differentiate between joint and barrel leaks. Requires launch access.	Successfully used in UK water industry. Used more for leak investigation rather than general survey work.
Temperature probes (Surface)	Long reach temperature probes to detect changes in ground temperature.	None identified.	Other techniques required to determine if joint or barrel leak.	No known usage.
Thermography (Surface/Aerial)	This is a well established technique for identifying temperature differences from an aerial survey and can show up surface temperature variations due to presence of water.	Long lengths of pipeline can be screened for leakage indications.	May not accurately pinpoint source of leakage. Other techniques required to determine if joint or barrel.	Currently used for leakage screening in the water industry. Other techniques may need to be employed to pinpoint source of leakage.
Tracer Gas (On-pipe/Surface)	This involves the injection of a tracer gas into the water system, which can then be detected at the surface of the ground.	None identified.	Limited success.	May have potential with Helium. Unknown usage.

JOINT REPAIR TECHNIQUES

This section provides a logical approach to the selection of joint repair techniques for any particular repair scheme. Figure 9.6 shows the steps that should be taken and the following sections describe the terms used and provide guidance on economic and technical considerations.

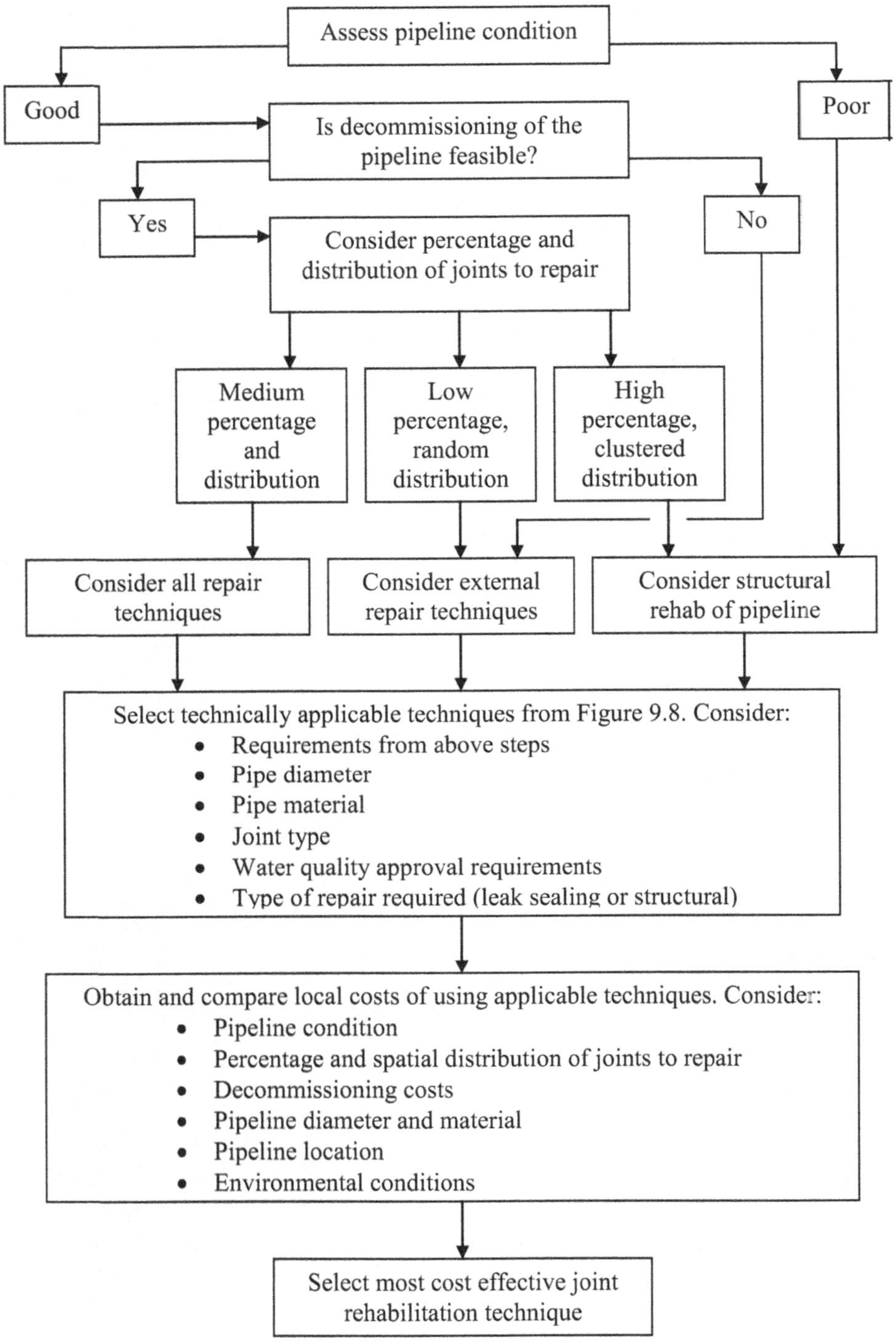

Figure 9.6 Selection of joint repair techniques

Economic Assessment of Joint Repair Techniques

The costs of joint repair are affected by many local factors. These factors can include; equipment and labour costs, environmental conditions, water quality and regulatory conditions, and availability of repair techniques and equipment. Costs can therefore vary widely throughout the world and it is not considered viable to give globally applicable cost information on joint repair techniques. This section provides a guide to the major parameters that must be considered in the selection of the most cost effective repair techniques for any particular repair scheme.

The economics of joint repair are affected by the following parameters:

- Pipeline condition
- Decommissioning
- Percentage and spatial distribution of joints to repair
- Pipeline diameter and material
- Pipeline location
- Environmental conditions

In practice, there is a degree of interaction between these parameters, which can vary between repair techniques. Assessment of the costs of any particular repair scheme must therefore consider all parameters collectively, such that the most cost effective repair techniques can be selected.

Assessment of Pipeline Condition

Is the pipeline in good enough condition to warrant repair of individual joints or is complete rehabilitation/replacement more cost effective?

Water companies use a wide range of techniques for determining the condition of their water pipelines. These are not the subject of this report, but can include: burst and leakage history, assessment of water quality problems and structural condition assessment. An initial assessment of pipeline condition is important, since, for example, it is unlikely that individual joint repair would be cost effective where a pipeline has a poor history of pipe barrel failure, or a short residual life (as determined through local condition assessment). In these cases, total fully structural rehabilitation of the main would normally be required. In some exceptional circumstances, e.g. where there is an operationally critical reason for not decommissioning, it may still be necessary to consider individual joint repair.

Where the choice between individual joint repair and total structural rehabilitation of the pipeline is not clear-cut, the condition of the main must still be considered to determine the long-term cost benefits of different rehabilitation techniques. For example, if after condition assessment a pipeline was found to have a large proportion of leaking joints, and a projected remaining life of 20 years (based on pipe barrel inspection), an external joint repair technique would correct the joint leakage problem, but may not affect the remaining life. A structural internal liner would also correct the joint leakage problem, and would provide the additional benefit of increase the remaining life (and hence asset value) of the pipeline. The long-term economic benefits of each applicable repair technique must therefore be considered in relation to any prevailing asset management strategy.

Decommissioning

The feasibility of decommissioning the pipeline should be considered. In certain cases, the cost of decommissioning or the impact on customer service may be such that decommissioning is not considered viable. For example, a pipeline may be critical to supply and to provide an alternative supply, or manage the disruption to customers caused by temporary loss of supply, might not be economic. In such cases, external repair techniques may be more desirable. It should be noted that some internal joint repair techniques can be deployed remotely, without decommissioning. These could also be considered.

Where decommissioning is feasible, the cost of decommissioning can significantly affect the overall cost of using some repair techniques. For example, the cost of structural internal lining would significantly increase if there were a large number of service connections present, whereas, the cost of joint excavation and external repair would not be affected by service connections.

Percentage and Spatial Distribution of Joints to Repair

The percentage of joints needing repair in a section of pipeline and the spatial distribution of these joints gives an indication of whether individual joint repair or lining techniques are likely to be most cost effective. The relationship between the percentage of total joints to be repaired in a section and the distribution of those joints along the section is shown in Figure 9.7.

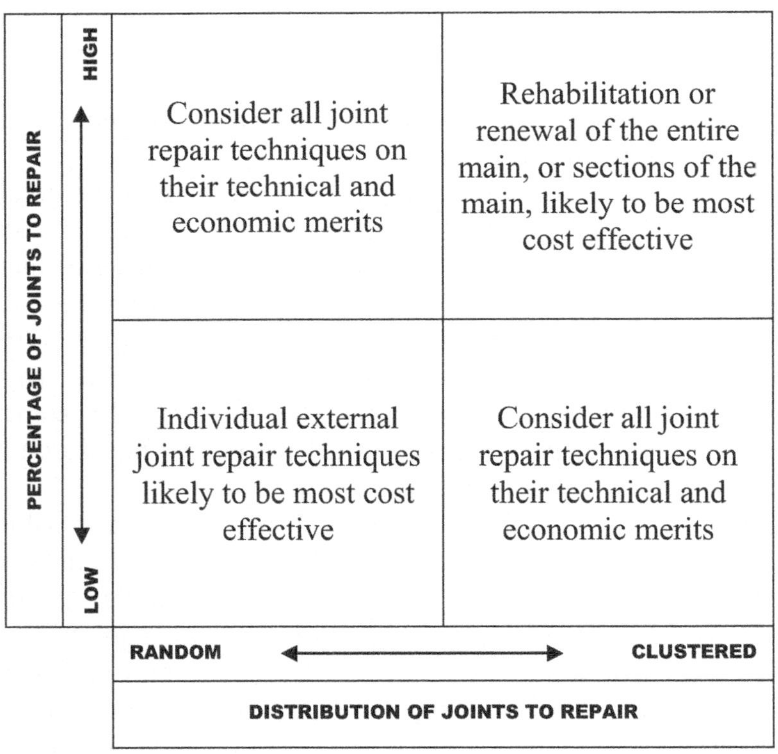

Figure 9.7 Relationship between percentage and distribution of joints to repair

As a general guide:

- A low percentage of defective joints randomly distributed along the pipeline will favour individual external repair, whilst;
- A high percentage or significant clusters of defective joints will favour internal lining techniques.

There is a large area between these extremes, where individual external repair and lining techniques should both be considered. This is likely to be the norm, and in these cases, applicable repair techniques should be selected through consideration of technical capability, as described in the following section "Technical Capability Assessment of Joint Repair Techniques". Local costs of using a range of applicable repair techniques should then be obtained and compared.

Pipeline Diameter and Material

These can have significant affects on the costs of consumables used in the repair process, and the cost of carrying out the repair process. Whilst costs increase with diameter for both individual external repair and internal lining techniques, they do not necessarily increase in the same proportions.

Pipeline Location

This has a major influence on the cost of excavation and reinstatement, traffic management and social costs. It is therefore very important to consider the location of pipes (major road, minor road, footpath, verge, open field, etc.) and to use this information when assessing the cost effectiveness of applicable joint repair techniques.

Environmental Conditions

Factors such as pipe burial depth and ground type should also be considered, since these can have a significant effect on excavation and reinstatement costs.

Technical Capability Assessment for Joint Repair Techniques

From initial consideration of pipeline condition, decommissioning feasibility and the percentage and distribution of joints to repair, it may be possible to decide on whether internal or external repair systems are most appropriate and whether leak sealing or fully structural repair is required, at an early stage in the repair technique selection process. The flow chart in Figure 9.6 provides guidance on this process. The repair requirements determined from this process can then be used to assist with selection of applicable repair techniques.

Figure 9.8 lists the joint repair techniques covered in this report, together with their key technical attributes. Table 9.4 provides descriptions of the techniques, together with benefits, limitations and notes on their suitability for use in potable water systems. Applicable techniques for a particular repair scheme can be selected from Figure 9.8, through consideration of the following attributes:

- Pipe diameter
- Pipe material
- Joint type
- Water quality approval
- Type of repair required (leak sealing or structural)

The first three of these attributes are self explanatory, allowing techniques to be selected based on the diameter, material and type of joints to be repaired. Note that generic materials and joint types have been used in the table.

Water Quality Approval

This refers to whether or not the rehabilitation technique is approved for use in potable water pipelines. This report covers a range of joint repair techniques, and whilst all of the techniques are suitable for joint repair, less than 50% are currently approved for use in potable water systems. Those not approved are currently used in other pipeline industries or are the subject of research and development. All of the techniques are potentially suitable for use in water pipelines. To provide an indication of current status, the repair techniques have been allocated status codes, as shown in Figure 9.8. Some water companies may wish to consider these potential techniques based on their technical merits, but subject to seeking approval for their use in potable water systems.

The Type of Repair Required

This is influenced by the condition of the pipe containing the joints as discussed above, but also depends on local conditions, and the reason why a joint needs to be repaired. Cause of failure and risk of further failures should be considered. For example, if joints were failing due to seal deterioration, then leak sealing techniques should be adequate. Whereas, if joints were failing due to deflection caused by ground movement, a fully structural repair may be preferred. In this context, leak sealing repairs can stop or prevent leakage from the joint and fully structural repairs are designed to withstand the same or greater internal and external loadings as the original pipeline.

Once available techniques have been selected from Figure 9.8, the economics of using these techniques should be considered. The section "Economic Assessment of Joint Repair Techniques" (above) provides guidance.

Joint Repair Technique	WQ approved	Internal or External	Structural repair	Status	Diameter range Min (inches)	Diameter range Max (inches)	Pipe Material Metallic	Pipe Material Plastic	Pipe Material Mineral based	Joint Type Push fit	Joint Type Bolted flange	Joint Type Welded	Joint Type EFC
Bolt replacement	Y	E	X	W, O	2	300	X				X		
CIP Liner	Y	I		W, O	3	60	X		X	X	X	X	
Clamps	Y	E	X	W, O	2	78	X			X	X	X	
Clock spring	N	E		P	4	56	X					X	
Close fit PE liner	Y	I	X	W, O	3	60	X			X	X	X	
Collars	Y	E		W, O	3	60	X		X	X			
Couplings	Y	E	X	W, O	2	96	X	X	X	X	X	X	X
Encapsulation	N	E		O, P	1	96	X			X	X	X	
External injection	N	E		O, P	3	48	X			X	X		
Joint sealing (packer)	N	I		S, O, P	4	118	X		X	X	X	X	
Local lining	N	I		S, P	6	30	X		X	X	X	X	
Pipe segments	N	I		S, P	18	196	X		X	X	X	X	
Reinforced cementitious	N	I		S, P	30	295	X		X	X	X	X	
Robotic repair	N	I		S, P	8	32			X	X			
Rubber seals	Y	I*		W, O	8	168	X		X	X	X		
Sliplining	Y	I	X	W, O	3	98	X		X	X	X	X	
Spirally wound	N	I		S, P	6	118	X		X	X	X	X	
Spray applied	Y	I		W, O	3	48	X		X	X	X	X	

* some systems can be applied remotely without decommissioning

Status Key	
W	Used in the water industry
O	Used in other industries; gas, oil, petrochemical
S	Used in sewage and waste water pipes
R	Subject of R&D for water industry use
P	Potential system

Figure 9.8 Technical attributes of joint repair techniques

Table 9.4
Summary of potential joint repair techniques

Joint Rehabilitation Technique	Description	Benefits	Limitations	Potential Suitability for use in the Repair of Joints in Water Distribution Pipelines
Bolt replacement	Systematic replacements of old and corroded bolts to maintain joint integrity.	Covers all joint sizes. Non-interruptive repair technique (although pressure reduction may be required). No capital equipment, minimum training and skill required.	Full excavation of each joint required. Corrosion of backing ring may be as significant.	Common method of water distribution pipe joint repair.
Clamps, collars	External methods of sealing a leaking joint by enclosing with a sealed mechanical fitting or assembling a new fitting to assist with the original joint seal.	Non-interruptive repair technique (although pressure reduction may be required). No capital equipment, minimum training and skill required.	Full excavation of pipe joint required. Specialist supply of collar may be required depending upon external diameter of pipe.	Common method of water distribution pipe joint repair.
Clockspring	External method of repairing leaking pipe defects using flexible sealing material to seal the leak and a composite overwrap to reinforce the damaged area	Quick, lightweight, minimum installation equipment required	Excavation of whole area to allow wrapping to be applied. Complex geometries may be more difficult	Utilised already for high pressure applications to seal leaking pipe barrels
Couplings	Mechanically jointed systems to repair a leaking joint by removing the whole pipe joint section and replacing with a new pipe length.	New fully pressure rated pipe section installed.	Full excavation of each pipe and interruption to supply required. Considerable undertaking for large diameter pipes. Specialist supply of couplers may be required depending upon external diameter of pipe.	Common method of water distribution pipe joint replacement.
Encapsulation	Mainly utilised in low pressure gas distribution operations to seal leaking joints using an externally applied thermosetting resin material.	Non-interruptive repair technique. Hand applied systems can accommodate complex joint shapes.	System relies on good adhesion to the pipe substrate. For higher pressure use, dedicated moulds must be fabricated to allow injection of the resin material.	Commonly used on gas and oil pipelines. Systems have been used on water distribution pipelines with fabricated metallic moulds, some materials are approved.
Internal rubber seals	Installation of an internal mechanical rubber seal at a leaking joint. Used extensively in man entry water mains.	Proven track record. Many joints treated from the same entry point in man entry sized mains.	Main requires decommissioning and internal access to the joints needs to be provided.	Used extensively in man entry water distribution pipelines and non-man entry sewer systems.

(continued)

Table 9.4 (continued)

Joint Rehabilitation Technique	Description	Benefits	Limitations	Potential Suitability for use in the Repair of Joints in Water Distribution Pipelines
Joint sealing (pig applicator)	The pig applicator is used to apply a volume of sealant material around the internal circumference of the pipe. Currently utilised extensively in sewer repairs.	Could be utilised in non-man entry mains. Several joints can be treated from one access point.	Main would require decommissioning. Internal access to the joints would need to be provided. Excessive levels of deposits would need to be removed. Pig applicator dedicated to a particular mains diameter.	Various forms have been used on gas and sewer pipelines, but sealant materials would require approval for use in water distribution pipelines.
Joint sealing (robotic)	Well-established technique to apply a sealant material to defective pipes and joints, generally in sewer networks. Some kind of remotely operated in-pipe vehicle is utilised.	Can be used in non man entry mains. Can be used to repair joints remotely from launch point. One tool usually accommodates several pipe diameters.	Main requires decommissioning and internal access to the joints needs to be provided. Equipment specialised and expensive.	Well established for sewer pipe renovation, some systems emerging for use in gas applications in non man entry mains. Sealants would require approval for potable water use.
Lining (cured in place) Patch repair for joints	The internal installation of sections of resin impregnated felt liners. The hoseliners are extensively used in sewers and some are available for potable water applications.	Could be utilised for non-man entry mains. Resistant to future joint movement. Liner can be tailored to provide the required level of structural integrity.	Main would require decommissioning. Internal access to the joints would need to be provided. Excessive levels of deposits would need to be removed	Currently used extensively on sewer pipelines and limited use for gas and water pipelines. A few systems have approval for potable water use.
Pipe segments	Internal insertion of short pipe sections (usually GRP or PVC) to ultimately build up a continuous liner for sewer pipe renovation.	New fully pressure rated pipe installed. Many joints replaced internally in one operation.	Main requires decommissioning, and new pipe tied into existing system. New service connections would need to be made.	Used extensively for man entry sewer pipe renovation however short sections across the pipe joints would require termination within the pipeline.
Reinforced cement	Spray or hand applied cementitious materials for the internal renovation of sewer pipes.	Well proven technique for pipe renovation. Many joints could be treated internally from one entry point.	Main requires decommissioning and internal access to the joints needs to be provided. Dedicated equipment required, with trained operators. Cleaning can be difficult and time consuming. Cure time of the liner material can be several days.	Used extensively for man entry sewer pipe renovation. Materials would need to be approved for use in potable water pipes.

(continued)

Table 9.4 (continued)

Joint Rehabilitation Technique	Description	Benefits	Limitations	Potential Suitability for use in the Repair of Joints in Water Distribution Pipelines
Remote internal rubber seals (AESL tool)	Prototype system for the remote installation of an internal rubber seal at a leaking pipe joint.	No interruption to the water supply. Can be used in non-man entry mains. System deployed remotely through a small tapping point. Several joints can be treated internally from one entry point.	Unproven prototype only available. Long term sealing properties currently unknown.	Concept based on use in non man entry water distribution pipelines.
Resin injection	External system to drill into the joint area (either in the joint face or through the back of the socket) and inject a sealant material directly into the joint.	Do not need to fully expose the whole joint or interrupt supply (pressure reduction could be required). Relatively small volumes of sealant are required.	Only currently designed to treat small low pressure leaks. May have to inject at several points around the joint circumference.	Successfully used on gas pipelines and often on flanged joints within chemical processing plant. Could be used for water distribution pipelines, but approval of the sealant material would be required. May be some limitation in the operating pressures for which the system can be successfully employed.
Shrink sleeves	Wraparound polymeric sleeve which shrinks in-situ upon the application of heat. Generally used to protect joints from further corrosion, but reinforced systems may provide additional support to prevent the joint leaking again in the future.	Flexible and versatile profile to accommodate joint geometry. Can accommodate variations in pipe external diameter.	Full excavation and cleaning of the joint required. Specialist training and equipment required. Specialist supply for larger diameter joints.	Commonly used as additional corrosion protection on metallic water distribution pipes. Limited pressure capability as a joint repair system.
Slip lining, Swageline, Rolldown, U-liner	Generally all well-established techniques for the internal installation of a continuous PE lining. These techniques should only be considered, if the pipe condition is poor and many joints are leaking.	Treats all leaking joints and a new, fully pressure rated pipe can be installed.	Main requires decommissioning, and new pipe tied into existing system. New service connections would need to be made.	Well established techniques for the rehabilitation or replacement of potable water pipes. Unless many joints are leaking, unlikely to be economic.

(continued)

Table 9.4 (continued)

Joint Rehabilitation Technique	Description	Benefits	Limitations	Potential Suitability for use in the Repair of Joints in Water Distribution Pipelines
Spray applied resin lining	Internal technique to spray apply a thin layer of resin to improve water quality in old cast iron pipes. Rapid cure system may allow several layers of resin to be applied depending upon the structural condition of the pipe.	Many joints treated internally from one entry point.	Main requires decommissioning and internal access to the joints needs to be provided. Dedicated equipment required, with trained operators. Cleaning can be difficult and time consuming. Cure time of the liner material can be several days.	Used extensively in potable water pipeline applications, but not specifically to treat leaking joints. New fast cure materials may allow a thicker lining layer to be applied to repair leaking joints.
Spirally wound pipe	A ribbed plastic strip which is spirally wound by a winding machine to form a liner which is inserted into the defective pipeline	Rapid insertion and can provide some structural support over long lengths	Access to the pipe required and only circular cross sections can be treated	Used in wastewater pipe applications, may be difficult to apply in short lengths.

CHAPTER 10
CONCLUSIONS AND RECOMMENDATIONS

CONCLUSIONS

The major findings are shown below and conclusions have also been drawn from each of the various sections within the report.

- Due to the variable nature of joint construction and as laid condition, the work undertaken has shown that it would be very difficult to provide measurable parameters and limits that would give a quantitative estimation of time to failure. The degree of leakage from a joint is currently the most reliable indicator of condition. However, the "acceptable" limits of leakage may still vary due to the operating constraints of the particular utility.
- The report suggests a number of parameters that could be measured or monitored to give an indication of joint condition, and the relative probability of future joint failure e.g. if a joint has deflected and become metal bound or spigot pull-out has reached unacceptable limits.
- Given the current availability of techniques for condition assessment of joints, it is considered that specific joint condition assessment may not be cost effective. However, if undertaken as part of a more general pipeline condition assessment programme, it may become more economically viable.

Joint Statistics

- There are a vast number of different joint types associated with the pipeline materials currently operating within water distribution networks.
- Based on installed pipe lengths and individual spool lengths per pipe material, an estimated joint population has been established. This population is in excess of 330 million joints in service in around one third of the potable water pipelines in the USA and Canada, along with all potable water pipelines in the UK and France.
- Data suggested that joint problems account for around 6% of all water distribution pipeline failures (excluding service connections) which require utility action.
- For all pipe diameters, due to the population of cast iron pipelines still in service, it is suggested that push-fit lead/yarn joints are still the most numerous.
- For larger pipe diameters (e.g. above 16 inch), data suggested that the concrete pressure pipe population, with its associated joints, is the most numerous in the USA.

Utility Survey

- It was evident that recording of joint failures (failure types and causes) has not been rigorously undertaken by water utilities in the past.
- The limited response to the utility survey means that although the data set is not statistically robust, it did provide some useful information:

- CI and/or ductile iron were generally dominant for all diameters, with local variations in other material types.
- For CI, DI and PVC, push-fit socket and spigot joints are the most common.
- For Steel and PE pipe, butt welded joints are the most common.
- CI showed the largest number of joint failures in a typical year, even for larger diameters.
- The dominant type of joint failure for all mechanically jointed pipes was gasket/seal failure followed by bolt failure.
- Socket failure of CI and PVC pipes was also significant.
- The primary cause of joint failure for mechanically jointed pipes was gasket degradation.
- The primary cause of joint failure for mechanically jointed non-metallic pipes was ground movement.
- Joint failures are mostly repaired by cut-out and replacement.

Understanding Joint Behaviour

- For the many different joint types in service, there are many common parameters, which contribute to joint degradation.
- An assessment of joint degradation can be undertaken by a ranking system based on leak classifications ranging from a non-leaking joint to a constant stream of water emanating from a joint.
- From the samples studied, iron socket and spigot joints are capable of performing satisfactorily with a wide variety of lead/yarn content, joint gap and angular displacement.
  - It seems impractical to develop techniques to measure the above parameters as they provide little information on condition or projected life before leakage occurs.
  - Monitoring changes in angular deflection or axial pull-out may provide an indication of joints that are most at risk of failure. This could be of particular use following seismic activity.
- The work undertaken on flange joints suggests that good gasket performance depends upon appropriately designed gasket, clean faces and the correct level of bolt torque to provide the necessary seating and sealing forces.
  - Once installed, there does not appear to be a readily accessible parameter that could be measured to provide an indication of bolted flange joint condition, other than that the rubber will degrade with time however:
    - on full faced flanges, an external ultrasonic technique may have the potential to assess the overall soundness of the joint.
    - access to components such as bolts can also provide useful information.
- From the tests performed on push-fit DI joints tests with elastomeric seals:
  - Small diameter pipes are more likely to fail in pure bending along the pipe barrel rather than crushing or bending failure at the joint.
  - Larger diameter pipes are more likely to fail in crushing adjacent to the joint rather than in pure bending.

- The tendency for the joint area to re-round as a result of angular displacement from increased load, therefore maintaining good seal contact suggests that angular deflection would not necessarily lead to loss of sealing performance.

- In mechanical testing of CI lead/yarn joints:
  - The first signs of leakage were generally between the spigot and the lead. If disbondment or the presence of water between the lead and the spigot could be detected, this would be a potential method of condition assessment.
  - In stepped tensile (axial) loading to failure, the lead will pull out of the socket (if an external raised step is present on the end of the spigot) or the spigot will pull out of the lead if no step is present.
- Potential measurement parameters indicating the onset of failure due to axial pull out: spigot withdrawal, lead withdrawal, disbonding of interface between lead and iron, leakage.
  - In cyclic tensile (axial) testing, leakage was generally worse during the compression part of the cycle.
- Potential measurement parameters indicating the onset of failure due to cyclic axial displacement: disbonding of interface between lead and iron, leakage.
  - In the bend tests, the observed leakage path was at 90 degrees to the plane of bending; i.e. at the expected points of rotation of the spigot in the socket.
- The location of weeps on an exposed joint may give an indication of the plane of bending and therefore be useful when assessing the cause of failure.
- Potential measurement parameters indicating the onset of failure due to bending: variation in joint gap around circumference, angular displacement, spigot and socket becoming metal bound, disbonding of the interface between lead and iron, leakage.

Finite Element Analysis Modelling

- Modelling failure modes of cast iron lead/yarn joints proved to be very complex and the limitations of the NASTRAN modelling package used were evident.
- The modelling results suggested that for cast iron lead/yarn joints:
  - Stress concentrations are dependent upon the loading direction, lead geometry, pipe fillet radii and the contact properties such as friction/adhesion.
  - The peak stresses are developed in the lead until a joint becomes metal bound, after which the contact area distributes high stresses between socket and spigot.
  - Material properties are not generally the dominant factors in joint behaviour. Interfacial characteristics such as adhesion appear to be more important.
- To build up confidence in the FEA modelling output, more accurate input data is required from laboratory testing and benchmarking.

Technology Survey

- A number of techniques were identified from the technology survey for the detection of joints in water distribution mains. These were categorised as surface applied techniques, techniques that require intimate contact with the pipe and techniques that are intrusive (surface, on-pipe, and in-pipe respectively).

  - Of these techniques, only CCTV is currently regularly used within the water industry, although there are a number of other techniques used in other utility industries (or under development), which offer potential in this application.
- Detection of joints does not always imply location of the joint. For example, if the objective was to detect joints so that external repair clamps could be fitted, in-pipe joint detection techniques would only be valid if they could provide an accurate above ground location so that the joints could be excavated and repaired.
- In condition assessment of joints, the level of deterioration could be assessed for certain parameters e.g. rubber hardness, if the original and current values could be compared or monitored periodically.
- The use of various ultrasonic techniques offers potential to obtain useful information on the condition of joints.
- There are few dedicated systems currently being used within the water industry to externally examine the integrity of non-leaking metallic joints.
- There are a number of technologies used in other utility industries (or under development), which offer potential in this application by measuring particular joint characteristics such as angular deflection.
  - There are systems to externally examine the integrity of non-leaking PE butt fusion and electrofusion joints starting to be used within the water industry.
- There are few dedicated systems currently being used within the water industry to internally examine the integrity of non-leaking metallic joints other than techniques such as CCTV inspection and pressure testing.
  - There are a number of technologies used in other utility industries (or under development), which offer potential in this application by measuring angular deflection, misalignment and axial separation.
- There are a number of leak detection techniques available which would be appropriate for joint condition assessment if it could be confirmed that the leak was from a joint and not the pipe barrel.
- If the condition of the joint can be assessed, it would be possible to select the most appropriate repair/renovation technique according to condition
- There are a number of techniques already available within the water industry for the external repair of individual pipe joints, and there are several others used by other utility industries, which offer potential in this application.
- There are a number of internally applied systems for the repair and renovation of individual joints within both potable water and wastewater pipelines.
  - There are a number of resin and mechanically applied systems used in other utility industries (or under development), which offer potential in this application.
- Several of the techniques considered are generally for pipeline rehabilitation rather than individual joint repair. However, in situations where there are a number of leaking joints, they may be the most appropriate option, especially for smaller diameter pipelines.
- A weighting system can be utilised by utilities as an aid to evaluating repair and rehabilitation techniques.

Decision Support Tool

- To assist water companies in obtaining the most appropriate technical, practical and economic solutions, a logical approach to the selection of techniques for joint detection, condition assessment and repair is necessary.
  - A decision support methodology provides a source of reference and a logical approach to technique selection.
  - Although the methodology stands alone, a supplementary CD is provided to assist water utilities with the selection of techniques.

RECOMMENDATIONS

- This report has addressed the AwwaRF project objectives through consideration of the technical performance of current and potential systems for joint detection, condition assessment and repair. However, it is recommended that the economics of joint condition assessment are investigated in more detail to determine the potential utility market and therefore the viability of further developing techniques specifically for this application.
- Areas prone to seismic activity could be a potential market that should be investigated further, and may prove to be more cost beneficial.
- It is recommended that a study is undertaken to determine what level and detail of information on joint condition would be required by a utility asset manager to trigger remedial action.
- Utility companies should be actively encouraged to record data on joint failures and the data should be gathered annually and compiled into a comprehensive database.
- AwwaRF should produce a standard data set for recording joint failures and circulate this around member companies.
- Collection and examination of more field joints would enable a better understanding of the construction and failure modes of old joints.
- Destructive testing of old and new joints is recommended to gain a better understanding of how joints fail, the parameters to be measured or monitored, and the further development of techniques to enable this to be achieved.
- Further pressurised DI bend tests are recommended to look at sealing and failure modes combined (in an appropriate rig) e.g. to confirm that failure occurs outside the joint area in bending.
- To undertake more extensive modelling, a software package more applicable to this application will be required.
- A comprehensive programme of work should be undertaken to gain a better appreciation of the required input parameters for FEA modelling, and further joint types should be considered e.g. PCCP.
- Further consideration should be given to some of the techniques for joint location and condition assessment used in other industries, and a detailed assessment of the potential market and development required should be undertaken.
- Ultrasonic techniques should be investigated further as they have shown potential in determining the integrity of:

- Rubber gaskets in flanged joints
- Lead/pipe disbondment in CI lead/yarn joints
- Fusion joints in plastic pipes

- Further consideration should be given to some of the techniques for joint repair and renovation used in other industries, and a detailed assessment of the potential market and development required should be undertaken.
- The decision support methodology should be used to assist with selection of techniques for joint detection, condition assessment and repair.
- Utility companies should be encouraged to provide feedback and additional local knowledge to enable the decision support tool to evolve.

REFERENCES

AESL, 2003, GB Patent Application Number. GB 0208318.6

Angus, H.T. 1976. *Cast Iron: Physical and Engineering Properties*. London: Butterworth's.

ASTM (American Society for Testing and Materials). 2002. *Standard Test Methods for Vulcanised Rubber and Thermoplastic Elastomers*, D412-98a. Philadelphia, Pa.: American Society for Testing and Materials.

ASTM (American Society for Testing and Materials). 1999. *Standard Test Method for Rubber Property – Pusey and Jones Indentation.* D531-89. Philadelphia, Pa.: American Society for Testing and Materials.

ASTM (American Society for Testing and Materials). 2001. *Standard Test Methods for Rubber Properties in Compression.* D575-91. Philadelphia, Pa.: American Society for Testing and Materials.

ASTM (American Society for Testing and Materials). 2001. *Standard Test Methods for Rubber Properties in Compression or Shear (Mechanical Oscillograph).* D945-92. Philadelphia, Pa.: American Society for Testing and Materials.

ASTM (American Society for Testing and Materials). 1964 (Withdrawn). *Method of Test for Indentation of Rubber by Means of a Durometer*. D676-59. . Philadelphia, Pa.: American Society for Testing and Materials.

Attewell, P.B., and J. Yates. 1986. *Soil Movements Induced by Tunnelling and their Effects on Pipeline Structures.*

Avon Lippiatt Hobbs Ltd., 1999, GB Patent No. GB 2184810.

AWWA (American Water Works Association). 1993. *Standard for the Design of Prestressed Concrete Cylinder Pipe*, ANSI/AWWA Standard C304-92. Denver, Colo.: AWWA.

Aziz, N.A., Finite Element Analysis of Push-Fit Elastomeric Spigot and Socket, MSc Thesis, University of Malaysia.

Bowden, F.P., & D. Tabor. 1964. *Friction and Lubrication of Solids*. Oxford: Clarendon Press.

Bridges D.Y., 1991, US Patent Number. US5076618.

BS (British Standard) BS78. 1917. Specification for Cast Iron Pipes and Special Castings for Water, Gas and Sewage. British Standards Institution. London.

Buchter, H.H. 1978. *Industrial Sealing Technology*.

Consumer Gas Co. Ltd., 1995, European Patent Number EP0660023.

Czernik, D.E. 1996. *Gaskets – Design, Selection and Testing.*

Drillflex FR, 1998, Patent Number WO098/59151.

Giller, G. et al. 2000. Technologies and Hardware for Ultrasonic Testing of Welded Joints of Steel and Polyethylene Pipelines. 15th World Conference NDT. Rome.

Hakko Co. Ltd, 1989, Japanese Patent Number JP1236975.

Heikki, L., Lembit, M., and Warwick, M., 1990, Patent Number WO9005267.

Horst, G.J.P., F.G.E.B. Weyn, and R.J.M Hermkens. Ultrasonic Inspection of Butt-Fused PE Joints, Gastec N.V., Netherlands.

Huebner, K.H., D.I. Dewhirst, D.E. Smith, & T.G. Byrom. 2001. *The Finite Element Method for Engineers*. New York: John Wiley & Sons.

Hunaidi and Giamou. 1998. Ground Penetrating Radar for Detection of Leaks in Buried plastic Water Distribution Pipes.

Jackson et al. 1997. An Evaluation of GPR Within The Water Industry.

Johnston Construction Ltd., 1982, US patent Number US347018.

Lee, J.M. 1997. MSC.NASTRAN Version 69+ Linear Static Analysis User's Guide, MSC Software, Los Angeles.
Lee, S.H. 1992. MSC.NASTRAN Handbook for Non-linear Analysis, v67, MSC Software, Los Angeles.
Lemaitre, J & Chaboche J.L. 1998. *Mechanics of Solid Materials*. Cambridge University Press.
Lide, D.R., ed. 1995. *CRC Handbook of Chemistry and Physics*, 76th Edition. Boca Raton, Florida. CRC Press.
LMK Enterprises, 1999, US Patent Numbers US5794663/5765597/6105619/6021815/6199591.
Messer, et al. March 2003. Novel High Resolution Defect Detection for Thermoplastic Butt Welds, Pipeline and Gas Journal.
Miller Pipeline Corporation, 2000, US Patent Number US6126206.
Neville, A.M., 1995. *Properties of Concrete*. Harlow, Essex: Prentice Hall.
NDTnet, April 1996, Vol. No.04, TWI, (www.ndt.net/article/twi/twi_1.htm).
NKK Corporation, 1990, Japanese Patent Number JP2203095.
Osaka Bodui Kensetsusha KK, 1982, Japanese Patent Number JP57057991.
Osaka Gas Company Limited, 1989, Japanese Patent Number JP1164890.
Owens Corning Fibreglass Corporation, 1986, US Patent Application US4589447.
Press, W.H., S.A. Teukolsky, W.T. Vetterling, & B.P. Flannery B.P. 1996. *Numerical Recipes in Fortran 77*. Cambridge University Press.
Rabinowicz, E, 1995. *Friction and Wear of Materials*. New York: Wiley.
Rad'ko, V.P., V.A. Troitskij, E.O. Paton E.O. 2000. Detection and Localizing of Defects in Welded Joints of Items From Plastic Using Ultrasound Electric Welding Institute of the NAS of Ukraine, 15th World Conference NDT, Rome.
Reiber, S. 1993. AwwaRF report - Chloramine Effects on Distribution System Materials.
Singal, A.C., and J.C. Benavides. 1983. Axial and Bending Behaviour of Pipeline Joints. J. Amer. Water Works Association 75, 572-578.
Smallman, R.E., & R.J. Bishop. 1999. *Modern Physical Metallurgy and Materials Engineering*. Oxford: Butterworth-Heinemann.
Smith, A.I., & A.M. Nicholson ed. 1971. *Creep Strains in Thick Rings Subjected to Internal Pressure, Ch. 17, Section III (Creep and Design of Components) of Advances in Creep Design (The AE Johnson Memorial Volume)*. London: Applied Science Publishers.
Sumitomo Metal Industries, 1989, Japanese Patent Number JP1040789.
Tokyo Gas Company (a), 1986, Japanese Patent Number, JP60139991.
Tokyo Gas Company (b), 1986, US Patent Number US5372162.
Toa Gurauto Kogyo KK, 1990, Japanese Patent JP2164952.
Toa Gurauto Kogyo KK, 1996, Japanese Patent JP8093044.
Tscheliesnig, P. et al. 1996. An Acoustic Leak Detection Pig, WCNDT.
Warnock, J.S. 1999. A Study of the Factors Influencing Mechanical Joint Performance in Water Pipelines. PhD Thesis, Imperial College, London University.
Willschuetz, K. D., 2000, Patent Number WO00539151.
Winiewicz, A. and D. Jarnecke, Nov 2002, The reference is associated with the New York Gas website sponsored projects -www.nygas.org/details/m-2002-013.htm
Wolf, Thomas. September 2001. Ultrasonic Inspection is Used Most to Detect and Size Cracks, Pipe Line & Gas Industry Features, Vol. 84 No.9. www.pipe-line.com/archive
WRc. (Water Research Centre). 1992. A Guidance Manual for the Structural Condition Assessment of Trunk Mains, WRc, Swindon.

Yedlitepe University. November 2003. Non Destructive Testing of Fusion Joints of Polyethylene Piping By Real Time Ultrasonic Imaging, 2nd International NDT Symposium & Exihibition, Istanbul, Turkey.

Zienkiewicz, O.C., & R.L. Taylor. 1989. *The Finite Element Method*. London: McGraw-Hill.

ABBREVIATIONS

AC	asbestos cement
ACPA	American Concrete Pressure Pipe Association
AESL	Advanced Engineering Solutions Limited
ASME	American Society of Mechanical Engineers
ASTM	American Society for Testing and Materials
Awwa	American Water Works Association
AwwaRF	Awwa Research Foundation
BS	British Standard
CAD	Canadian dollar
CCD	charged coupled devices
CCTV	close circuit television
CI	cast iron
CIPP	cured in place pipe
CPP	concrete pressure pipe
DGPS	differential global positioning system
DI	ductile iron
DN	nominal diameter
EF	Electrofusion
EPDM	Ethylene Propylene Diene Monomer
ESI	Engineering Services Incorporated
FEA	Finite element analysis
GIS	Geographical information system
GPR	ground-penetrating radar
GPS	global positioning system
GRP	glass reinforced plastic
GUI	graphic user interface
GUL	Guided Ultrasonics Limited
HARC	Houston Advanced Research Centre
HDPE	high density polyethylene
in.	inch
INS	inertial navigation system
IR	infra-red
kg	kilograms
kN	kiloNewtons
KNDT&I	Korea Non-destructive Testing and Internet Company Ltd.

lbs	pounds
LCD	liquid crystal display
LDPE	low density polyethylene
LVDT	linear variable displacement transducer
MDPE	medium density polyethylene
MFL	magnetic flux leakage
MIG	metal inert gas
mm	millimetres
MPa	megapascals
MRS	minimum required strength
MSC	Macneal Schwendler Corporation
mS/m	milliSiemens/metre
mV	milliVolts
NRC	National Research Council
NWL	Northumbrian Water Limited
NWTC	Northumbrian Water Technical Centre
NYSEG	New York State Energy and Gas Company
PCCP	pre-stressed concrete cylinder pipe
PE	polyethylene
PE EFC	PE electrofusion coupler
ppt	parts per thousand
PSC	pre-stressed concrete
psi	pounds/in^2
PTFE	polytetrafluroethylene
PVC-U	unplasticised polyvinyl chloride
PVC-O	molecular orientated PVC
PVC-A	modified PVC
PZT	lead zirconium titanate
QA	quality assurance
RC	reinforced concrete
RFEC	remote-field eddy current
RSG	resistance strain gauge
SDR	standard dimension ratio
SDRC	Structural Dynamics Research Corporation
S Test	Structural Technologies Limited
TIG	tungsten inert gas
TOFD	time of flight diffraction
TRRL	Transport and Road Research Laboratory

TWI	The Welding Institute
UK	United Kingdom
UKWIR	UK Water Industry Research Limited
US	ultrasonics
USA	United States of America
USD	US dollar
UTS	ultimate tensile strength
WIS	Water Industry Standard
WQ	water quality
WRc	Water Research Centre

IWA Publishing's authorised EU representative for General Product Safety Regulations is Diane D'Arras, 15 rue Duret, 75116 Paris, France, e-mail: safety@iwap.co.uk.

Printed and bound by CPI Group (UK) Ltd, Croydon, CR0 4YY
23/09/2026
02221720-0002